RFD

RFD

Charles Allen Smart

Foreword by Gene Logsdon

Ohio University Press   Athens

Ohio University Press, Athens, Ohio 45701
© 1938 by W. W. Norton & Company, Inc.
Foreword © 1998 by Gene Logsdon
Printed in the United States of America
All rights reserved

First Ohio University Press edition published in 1998.
Originally published as *R.F.D.* by W. W. Norton & Company in 1938.
Reprinted by arrangement with W. W. Norton & Company, Inc.

Ohio University Press books are printed on acid-free paper ⊛ ™

02 01 00 99 98   5 4 3 2 1

Frontispiece: Charles Allen Smart, in a photo at the time of *RFD*'s first
printing. *From Charles Allen Smart papers, courtesy Ohio University Libraries*

Book design by Chiquita Babb

Library of Congress Cataloging-in-Publication Data
Smart, Charles Allen, 1904–1967.
    RFD / by Charles Allen Smart ; foreword by Gene Logsdon.
        p.   cm.
    Originally published: R.F.D. New York : W.W. Norton, © 1938.
    ISBN 0-8214-1254-X (pbk. : alk. paper)
    1. Farm life—Ohio—Chillicothe. 2. Smart, Charles Allen,
    1904–1967. I. Title.
    S521.5.03S535   1998
    630'.9771'82—dc21
    [b]   98-24825

To

Douglas Auld Shepardson

Contents

Foreword

Before I started reading this book, I had made up my mind
not to like it. Charles Smart was just about everything I was not
or didn't altogether trust: an urban intellectual born to and married into rather wealthy families that kept him at a safe distance
from the dire claws of poverty that have always hovered threat-
eningly over me. He was a graduate of Harvard, oh God; in-
structor at Choate, oh God again; a romantic wanderer around
Europe hoping he looked like the novelist he was trying to become. Still, I could handle all that easily enough—*somebody*
must go to Harvard and teach at Choate and wander around
Europe. But then he had what seemed to me the hypocritical
effrontery to declare himself a socialist while inheriting two capitalistic farms in Ohio and a nice chunk of capitalistic money—
in the depth of the Great Depression when even a small inheritance was truly golden. To make matters a hundred times
worse (in my mind), he then decided to try play-farming on his
ill-gotten land, aided and abetted by what amounted to a small
but guaranteed annual income from the rent on the second
farm and returns on investments from his inherited money. I
started reading RFD with a hundred chips on my shoulders and
pen sharpened to a shark tooth's edge with which to rip this
precious fake gleefully to shreds.

A funny thing happened. I agreed with almost everything he wrote. More than that, I came to wish I had known him. I am certain we would have become close friends.

The main reason I'm sure of that, and the main reason, I think, that *RFD* became a best-seller when it was first published in 1938, is that Charles Smart was one honest man, something that is hard to find anymore in nonfiction books, especially about rural life. (You only get the truth in fiction, strange as that may seem.) He confronts my mistrust of his background head on, discussing his inheritance in detail and worrying exceedingly about the injustices inherent in a system that not only allows the passage of unearned wealth from one generation to another, but the impossibility of avoiding such inheritance even if he wanted to. And as it turns out, the amount of money in question was far from a fortune, just enough to tide him over when neither farm nor his writing earned much.

He frankly admits his lack of farming expertise over and over again, although it becomes clear that he knows quite a bit more about farming than many "real" farmers I could mention. He had spent the summers of his boyhood on this farm, and says that his life in those days "began and ended with summer vacation." He makes no bones about his seemingly radical socialism. "I am a fairly typical urban intellectual and malcontent of the twenties," he writes. "I am a rather ordinary young man in his thirties who after printing, publishing, literary hackwork, novel-writing, wandering, loafing, and teaching, went to live on a farm, and got married." But his "socialism," like Scott Nearing's, looks to me a whole lot more like true capitalism as he struggles to make a little money from his farm. In fact, compared to the rotten welfare capitalism of today, where megabusiness and big government divide up the public pie for their own benefit, Mr. Smart was an Ayn Rand. He finally abandons his flirtation with American Marxism, as Scott Nearing did, realizing that when it comes to reforming American agriculture, his farming neighbors are more trustworthy and knowledgeable than any institution.

He also finds more to admire in farmers than he does in his urban intellectual friends who come to visit him occasionally. Bowing to the complexities of farming, he says: "In my first three weeks in the publishing business, and my first three weeks as a teacher, I learned more about publishing and teaching than I have learned about farming in my first three years on a farm." And again: "I have learned more from my neighbors than they will ever learn from me . . ."

Mr. Smart was, to my way of thinking, the almost perfect kind of personality to put down roots in a rural community. Not only did he have the proper respect for the people already there, but he could gently point out the limitations of their cultural attitudes—the sometimes narrow-minded anti-intellectualism and puritanical religiosity that rural isolation (or city isolation) can produce. An infusion of his more easy-going and fun-loving renaissance philosophy was exactly what rural communities needed—and still need. If farmers were happier, really enjoying what they were doing, Mr. Smart believed, they would be less prone to a rigid, unbending work ethic that sacrificed everything for money—less prone to the constant expansion that drove their neighbors out of farming.

Mr. Smart's lifestyle is what you get when you cross an Amish farmer with a romantic Epicurean. He sings Gregorian chant while he milks his cow. (My grandfather did the same thing.) He writes: "I like to look at landscapes, thinking of paintings I have seen or may see. If more farmers could do this, they would spend . . . 20 percent of their time [in this kind of activity] . . . if they only had . . . the sweet craziness in the head, they might be . . . the happiest men on earth." He then adds, in humor, that such a trait, reducing so-called efficiency, could help end the overproduction that was depressing prices, and so gazing at landscapes would become quite profitable. Or again: "One advantage of bathing on the lawn is that the water is so easy to get, and so easy to dispose of. Another is that one can splash and spill and pour with abandon. The most important is that it is such pure sensual fun to be naked and bathe in the open air, on the green grass, under the trees, with gentle airs, and slanting sunlight, or even moonlight, caressing our bodies." Or again: "Good farming is more complex, even, I think, than good juggling or good painting, and the odds against perfection are even greater. I have never seen a farm that justified a sour face, and too many good farmers go around looking as though they were attending their own funerals, as indeed they are." And again: "Living in the country demands the most ardent and happy intellectual life of which one is capable." And again (I can't resist): "I don't think that any of us can afford to look at nature and at the major facts of the human situation while dead sober. Then they seem serious, irrational, and rather dreadful. When one looks at them in a very mild state of intoxication, they seem equally irrational, but more strange and amusing than horrible. I suspect, furthermore, that the slightly drunken view is the true one, more true simply because perceived with greater sympathy."

RFD is not, obviously, your standard how-to book about moving back to the land. Of course, like all good books on rural living, it contains more real how-to than all the how-to books on the subject put together, but that is not the first or the last aim of the book. What Smart presents here, rather, is a detailed description of life on an Ohio farm in the 1930s, lived by himself and his wife, combining urban romanticism and very down-to-earth practical farming. Along the way, Mr. Smart observes a great many things about human nature that give the book universality and timelessness. Almost by indirection, he takes the reader through all the multitudinous events of a general livestock and grain and garden farm through the seasons, leaving no doubt that while he may be "only" an urban intellectual, he has become also an astute farmer, by love and by economic interest. He goes into great detail about the money he makes and does not make but then puts the whole impossible discussion about money-making (in farming or anything else) on its proper footing with this perfect remark: "I trust the instinct that makes it easier to sleep with a woman, or to learn a man's most intimate ideas, than to learn the essential facts and figures of his of her private economy."

He pulls no punches about country living and cautions readers to think in terms of a little income, rather than to think they can make an entire living from a small farm. "I am now in a position to make with assurance my warning that small farmers, and especially small farmers from the city, simply must have a 'cash crop,' a trained and marketable ability, a specialty." And again: "The only people I know who make what educated and lucky city people would call a decent living have farms of a thousand acres or more [even then!]. My kind of farming is not for incompetents and weaklings; it is for people who have no interest in 'getting ahead,' who like animals and plants more than machinery, processes more than figures, solitude more than most company, and a hunting cap more that a derby hat." Ah, Charles, we would have been such good friends.

The reader learns that while rewards might come from farming, the farmer pays his dues in a life often arduous, physically uncomfortable, and mentally stressful, like every other way of making money from manual labor. But he also shows in a hundred ways why "even the poorest and most ignorant of them probably get more fun out of their work, itself, than most city workers get out of theirs." Or as a farmer friend of mine says: "The worst day on the farm is better than the best day in the factory." To the urban sophisticate (or farmer) who thinks rural life is boring, Mr. Smart writes: "Boredom is a mysterious matter, more appropriately handled by physicians and psychologists than by me. In myself, I find it caused less by the immediate situation, less even by more remote and less obvious circumstances, than by some obscure inner weakness, fatigue, and dullness in myself, or simply by something in the air pressure and weather. However it seems to me that it would take the most extreme and outrageous poverty, or wealth, or weakness, or lack of curiosity, or ineptitude for reading, watching, and conversation, to make country life dull. There is too much to learn."

At times Mr. Smart sounds eerily modern. Describing Chillicothe and its environs, where his farm was located, he writes: "There still seem to be some crafts and trades to which a young man can be apprenticed, and in which he has some chance of making his own way. There are still a few independent stores that make money, and by the time they are driven out, the chains may have become humanized." (Well, he can't be right all the time.) "Entirely too many people have jobs that are both unsuitable to them and precarious. Trade unions are feeble, or crooked, or both. . . . Plenty of people are strangled by installment buying. The equipment of the public schools is good, but the old irrelevant curricula endure, the football field seems to be considered generally more important than the library, and there is a multitude of silly 'activities.' People generally seem to be increasingly dependent on pleasures that have to be bought, like cars and movies, and therefore on money and on their employers. The government is expensive and inefficient, social services are rudimentary, the public library is neglected, and religion is competitive and social."

In sensing the future, Mr. Smart's most accurate observation about farming has become the battle cry of the new niche farmers of today who see a future selling direct to the consumer. In a section headed "Come and Get It!" Mr. Smart grows almost obstreperous in a way not characteristic of the rest of the book. Trying to sell the production of his farm as much as possible directly to the people who were going to eat it, he writes:

> When I sell, I don't have to "create a demand" by appeals to fear and vanity, and by an elaborate structure of lies, and my personality doesn't enter into it. . . . If you city folks can't buy my food and wool until it has been bought and sold a dozen times, and can't afford it then because you merely work, and don't own the factories, why whose fault is it? Come and get it, and I for one will meet you halfway.
>
> . . . You will have good food like mine, I shall have good plumbing like some of yours, and everyone will have a job. Everyone, that is except the buyers-and-sellers who don't know how to do anything else and who then, by God, will have a chance to "sell themselves," if they can.
>
> I speak with some heat, and I mean what I say, because every time I go down to the stockyards with something to sell, I see better men than I am who have nothing but their livestock to sell, and who haven't a chance in the world to work up a "specialty" like I know to my sorrow is absolutely necessary, right now, for decent life in the country. They haven't any typewriters and they haven't any papers in safe-deposit boxes. And I'm not just being altruistic and sentimental, either; I am too close to their situation, myself, for any of that. We want to earn a decent living for goods produced, and we want to get our meat and wool to you, and to you alone, for whatever it is worth to you.
>
> When are you going to come and get it?

Unfortunately in the nineteen-thirties not enough people resettled into rural areas and followed his lifestyle to stem the tide of rural outmigration and the ruthless, profiteering agribusiness that left in its wake only a land empty of community. Perhaps the new renaissance farmers now moving out into the forsaken countryside will make of Mr. Smart a culture hero yet.

And he should be considered a hero of some sort. Mr. Smart did not return to the city after his book became a best-seller, as I feared, but continued farming until 1942, when he joined the navy and saw action in the South Pacific even though at his age and with his farming occupation, he might easily have gotten a deferment. He came back to the farm, continued to work and write there and to teach at nearby Ohio University, and except for forays into Mexico a part of the time in later years, lived and died at his beloved Oak Hill Farm.

Gene Logsdon

Preface

This book is intended to be a picture of life on a farm in southern Ohio in the nineteen-thirties.

As a writer temporarily absorbed in a personal adventure, I have had to use the handiest models that were of interest to me; therefore, the farm is mine, and the central human figures are my wife and myself. I have had to report some of my own inner experiences, and to guess about the inner experiences of others. These facts need not, I think, make this book any more embarrassing than self-portraits and portraits of their wives by painters who couldn't afford other models. All the other little portraits I have introduced are from life, too, but they are only a part of one person's response to the people pictured, and have no pretensions to more complete and final truth. I don't think I have introduced anyone without trying to go deep enough to be essentially sympathetic. No injury is intended, and I hope none will be felt. All proper names of people have been changed, and other details have been altered to respect privacy.

But this book is not primarily autobiographical. I have introduced the past only when it seemed an element in the present, and I do not consider my character, experiences, and opinions, as those of an individual, so peculiar and important as to be worthy of autobiographical treatment. I am a rather ordinary young man, in his thirties, who, after printing, publishing, literary hack-work, novel-writing, wandering, loafing, and teaching, went to live on a farm, and got married. I was a fairly typical, urban intellectual and malcontent of the twenties, and I am now merely doing those things that millions of other Americans are doing without fuss and feathers, and without writing books about themselves. Whatever interest this book has will arise from the conjunction of this past with this present, and from the other lives with which my own is involved. As an individual, I am insignificant, but as a city-bred man, trying to get a living, in all senses, from the soil, I have found in myself, with these other figures, and against this background, enough interest and comedy to see me through a couple of years of hard and happy work as a writer.

I feel very grateful to many people, among the quick and the dead, but the living among them, and other readers, can find acknowledgments, explicit and implicit, elsewhere in the text. In addition, there are other debts less obvious. Once more I am indebted to my friend, Mr. George Stevens, for encouragement in a pinch, and for detailed criticism of the very highest quality. I am also indebted to my friend, Mr. Eugene D. Rigney, for much valuable assistance of several kinds. Our County Agricultural Agent, Mr. Fred R. Keeler, always a stout ally, has made valuable suggestions for which I am most grateful, but neither he nor anyone else but myself is responsible for my facts and opinions. Mr. Archibald MacLeish and his publishers, Messrs. Houghton, Mifflin Co., and Messrs. Farrar & Rinehart, have kindly allowed me to use a few lines from his poems.

I. Home

An American landscape.

The look and feel of a landscape can tell a great deal about the people in it, but this landscape in southern Ohio calls for painters and poets, and not for prose-men. I have seen an exhibition of the works of Ohio painters, but for the most part they were amateurish, and merely pretty. The only paintings I know that catch the essence of this landscape are little known, the work of the late H. H. Bennett, an editor, scholar, painter, and poet whom this region misses sorely, and too unconsciously. There are passages in the poems of A. MacLeish and S. V. Benét that suggest what I want, but are only tantalizing.

The hills of Ross County are really small and large plateaus, cut by gorges and very narrow to broad valleys. There are poor farms like this one in the highlands, and excellent farms along the valleys of the Scioto River, Paint Creek, North Fork, and other streams. The hillsides are heavily wooded with oaks, black walnuts, hickories, beeches, and ashes. There are some cedars, but almost no pines, and my wife misses keenly the evergreens along the coast of Massachusetts. Along the streams there are great sycamores, willows, and cottonwoods. In the bottoms there are large fields of corn, wheat, and hay, with barnyards full of cattle and hogs. In the hills there are scraggly patches of corn and vegetables, a few good orchards, and little shacks and cabins where people seem to hang on to life by their eyelids, die of rickets and typhoid, and shoot each other for adultery.

In the autumn, there are the rich reds, browns, and blacks of the woods and fields, the blue-gray of the hills, and the rather melodramatic effects of the autumn sky. There is also the rich green of the winter wheat, like a carpet beneath the fading shocks of corn fodder, with thousands of little piles of corn, pure gold, that has been husked, but not yet hauled in. In gardens, the weed-piles are smoking slowly, and woodpeckers rattle the seed-pods of the yuccas.

Most men wear hunting caps, and one hears them in the woods—the pure, lovely, and harrowing baying of the hounds, the reports of rifles and shotguns, and sometimes the whine of a bullet. At night one meets the hunting parties on the roads, with lanterns, dogs, and guns. The talk in stores, bars, filling stations, and grain elevators is all about pheasants, rabbits, squirrels, coons, and partridges, and men look at me askance when they discover I have no stomach for hunting. It is all normal and well controlled, and often the food is needed. It could even be called humane: for me, people come first. I certainly do not raise my hands in horror. I just could not shoot a bird, or anything else, and if I did, I'd not get over it for a week. For me, there is enough death abroad as it is, thank you. But hunting is only an element in the autumn landscape, and not an inappropriate one. Everything else is dying, and being tidied up; one breathes in the first chill in the air and wonders where one is going.

Ordinarily, they say, winters here are mild, and this last one was; just more cold, wet, silence, mud, quiet work, and sleep. But the last two were not like this; they were fights to the finish, with worthy settings. Snow, and then more snow, with the trees stark, delicate, and the hills looming into a sinister grandeur. The winds rise quickly, from almost absolute silence, cut one to the marrow, and howl in the chimneys and around the eaves.

There are a few gallant little winter birds, twittering in the deserted garden and robbing the corn crib. The cattle and sheep huddle together in their shelters, or wander restlessly in the snow, bawling and bleating with hunger and cold. The dogs frolic for a while in the snow, and are glad to come indoors. The days are terribly short, with the sun sometimes staying hidden all day, and then blazing out on the bright desolation, and going down in taunting crimson glory. But most of the days are dull gray, with night falling like early death. It is the end of the world: towards the end of winter, one begins to lose any real, emotional belief in the return of life.

And then, suddenly, a warm day, and another, and everything begins to breathe, melt, and tremble. Winter has sharp teeth, still, but something has broken. One day at the end of February my hired man and I were sawing away at a huge oak log, and we paused for breath, and heard, deep in that log, a crack!—like a bell of liberty. March is wild and treacherous, of course, a big flamboyant blonde coquette, windy with passion and laughter, excellent gay company, grand and mean, and treacherous as the foul fiend. And then April, gentle and innocent, green and sad and puzzled as all youth, with quiet rains, and freshness everywhere, and moments of cold fear and spite. And then with May, everything bursts into green in the sunlight, the days are long with good work and rest, the birds and lambs are a bit hysterical, the dogs and cattle find their old haunts in sunlight and shade, the gardens begin to look like something, the pastures are full of colts, and life goes on.

June, July, August, and September, deep summer, with everything reaching its limits in growth, and work, and play, and the land a deep and quiet place, green, wet, dark, and cool, or hot and dusty, with towering clouds and sudden lightning, and everything slowly growing a little wise and a little tired, is as quiet, dull, happy, strong, and deceiving in time, as middle age—and (damn it) I like it, best of all.

Sometimes in this country I see hints of other places I have seen and loved. From a hill road, now and then, one sees a bit of rolling country, with tall woods, that is like nothing but the Cotswold Hills of England. Or an old house, buried in dripping green in spring, that is like a house seen once in Wiltshire. Or a solitary elm, dreaming delicately above a stony field, that is New England. Or a rocky shoulder, with a vineyard, and great trees, that means Burgundy. Or the broad valley of the Scioto, as it nears its mouth in the Ohio, that suggests vaguely the lower reaches of the Rhône. But these little hints and suggestions are all momentary and false. There is nothing else in the world like this little Ross County, this Ohio, this America; there is nothing else like it at all.

This is what A. MacLeish was talking about, I think, when he wrote, in "Landscape as a Nude":

> Under her knees there is no green lawn of the Florentines:
> Under her dusty knees is the corn stubble:
> Her belly is flecked with the flickering light of the corn:
>
> She lies on her left side her flank golden:
> Her hair is burned black with the strong sun:
>
> The scent of her hair is of dust and of smoke on her shoulders:
> She has brown breasts and the mouth of no other country:

With people, buildings.

I hesitate to say what kind of people and buildings there are in this landscape that I love. I am in some ways, still, an ignorant outsider, such realities are extremely complex, and I have some of the dangerous book-ideas and temptations to simplify hastily that are characteristic of literary men. If you are interested in the Middle West, or in any other part of America, my advice is to distrust novelists, lecturers, literary revolutionaries and reactionaries, and the millions of living Carol Kennicotts. Go rather to the historians and sociologists, to good poets, and, with a good supply of salt, to the movies and popular magazines. I have to use books as tools, and those that I have found most helpful, in my attempt to understand our nearest town, Chillicothe, and Ross County, Ohio, are *Middletown* and *Middletown in Transition,* by Robert S. Lynd and Helen Merrell Lynd, and *Technics and Civilization,* by Lewis Mumford.

Chillicothe itself, a city of nearly twenty-five thousand, is not, I suppose, much different from a thousand other cities of the size between the Appalachians and the Rockies. It has a couple of large paper mills, a shoe factory, railroad shops, lumber yards, grain elevators, livestock sale barns and yards, chain and independent stores, banks, garages, nurseries, federal industrial reformatory and veterans' hospital, a great many ugly houses of all periods and styles, a few nice old houses and trees, a country club, tourist homes, pool rooms, small restaurants and lunch stands, small, mediocre hotels, filling stations, three movie theaters, a few old business blocks, many churches, whore houses, beauty parlors, courthouse, city hall with police station, post office, fire houses, public schools, parks, library, a few rudimentary suburbs, a reservoir full of hard water, hospital privately supported, cemeteries, and all the rest of the familiar plant and equipment of so-called civilization in the Middle West of the U.S.A. in the nineteen-thirties.

On a smoky autumn Saturday evening, with the farmers pouring into town, the spectacle has great interest and some charm. Whatever else you can say about them, these people have been doing things with their hands and brains, most of them seem to know each other, and no one has ever read the Boston *Evening Transcript.* On a Sunday afternoon, however, no town this side of hell could be more cold, stark, and ugly.

Chillicothe isn't so bad as Middletown, because it is smaller, older, less industrial, and more individual. In some respects we are moving from what Mumford calls the Eotechnic Era, roughly of the seventeenth century in Europe, and the eighteenth and early nineteenth centuries in America, into the Neotechnic of the late twentieth, avoiding at least a few of the evils of the Paleotechnic of the late nineteenth. Except in the paper and shoe factories, which do not dominate the town, there seem to be fewer strictly mechanical and senseless jobs. There still seem to be some crafts and trades to which a young man can be apprenticed, and in which he has some chance of making his own way. There are still a few independent stores that make money, and by the time they are driven out, the chains may have become humanized. We are far enough south so that few people, except in one of the paper mills, are working themselves into neuroses. We are far enough north so that we are not completely dependent on Negroes, who are separate from us rather than subservient to us. I don't think that manual workers feel inferior to business men, or superior to them, or try to ape them. I could go on for some time, without flattering Chillicothe, in pointing out some of the ways in which it has escaped the evils of our time.

But on the whole, the picture is only a few shades less gray and dismal than that drawn by the Lynds. Unemployment is still serious. Entirely too many people have jobs that are both unsuitable to them and precarious. Trade unions are feeble, or crooked, or both. Rents are high, and on the whole, the best livestock in this county is housed more safely and appropriately than many of the people. Merely from looking at people in the streets and in the poorer homes, I should say that the public health situation was bad. Plenty of people are strangled by installment buying. The equipment of the public schools is good, but the old irrelevant curricula endure, the football field seems to be considered generally more important than the library, and there is a multitude of silly "activities." People generally seem to be increasingly dependent on pleasures that have to be bought, like cars and movies, and therefore on money and on their employers. The government is expensive and inefficient, social services are rudimentary, the public library is neglected, and religion is competitive and social.

In the country, the spectacle is little more heartening. In general, the resources in land have not been conserved. The County Agricultural Agent and enlightened farmers are on the job, but I'm afraid that payments under the Conservation Act will never effect the needed reforms. This region suffers severely from droughts and floods, and our only hope was in the Scioto-Sandusky Water Conservancy Project, which has been delayed, according to *rumor*, by state politicians and by army engineers, who have wasted colossal sums of money elsewhere in their hope to make it possible for warships to pass from New Orleans to the Great Lakes. (This sounds fantastic only to those who have never encountered the so-called military mind.) Farm tenancy is more humane here than in the South, but it is serious enough.*

Electricity, gasoline, and plumbing have lightened the work of those who can afford them, but few can. Our electric company is of the familiar type. Rates have been reduced by threats,

*The farm population of the county is about 20,000. Acres in farms, in 1920: 384,328; 1925: 371,189; 1930: 348,944; 1935: 387,783. This last figure divided as follows: cropped, 160,659; plowable pasture, 68,490; other pasture, 34,957; pastured woods, 50,256; not pastured woods, 54,483; other, 18,938. In 1924, 31% of the farms, and 43.5% of the land were operated by tenants; in 1929, 33% and 47%; in 1934, 32.1% and 46.3%. In 1929, farms operated by tenants averaged 163 acres, valued at $43.65 per acre; farms operated by owners averaged 115 acres, valued at $40.33 per acre. Of 370,993 acres m use in 1935, 60,224 acres, or about 16.2% have been recommended for retirement to unpastured woodland or permanent pasture. The gross farm income for 1929 was estimated at $2,288,757.04. I have no means of breaking this figure down with any confidence, as between landlords, owner-operators, tenants, and hired help, but if there were 4,000 farm families, they grossed $572. per fainily. It is estimated that if conservation is effective the gross will be raised almost half a million by 1947.

but I imagine they could go much lower. In the last few years, despite the weather, our farmers have had a better chance than at any time since 1919, but the basic situation is still, I think, very unsound. Living and working conditions in the hills still astonish and depress every visitor who sees what he can of them. I would rather live in one of these shacks than in the mill towns of Pennsylvania, or the match-box suburbs of Long Island, but that preference is probably quite personal and subjective. The farmers here have less time, energy, and money for fun than the people in town, but they seem to be more capable of amusing themselves. Farm boys are still eager to try their luck in town, but now, possibly, more of them return.

The most obvious social phenomena in this region are the two federal institutions north of town, on the site of a former army cantonment, Camp Sherman. I have never visited the Veterans' Hospital, but I have never passed it in a car without a new shudder at the waste of war, and without a reflective memory that this hospital was on the preferred list of the American Medical Association, while our city hospital was not. I have visited the Industrial Reformatory, or prison, and been much impressed. It is the best federal prison, and to it are sent the less dangerous convicts. The institution is a trade school, and the purpose is the reconstruction of the individual, in mental and physical health, skill, and basic attitudes, rather than punishment. The whole place seemed notable for its humanity, scientific attention to detail, and good sense. What depresses me is the idea that the inmates could never have had equivalent opportunities before becoming criminals, and that plenty of young men in this very county could begin their education best by stealing an automobile. Naturally, any life outside those walls is better than any inside them, but I should think a string of open schools as good as this one could be paid for easily out of the economies resulting from a reduction of crime. Good construction is always cheaper than reconstruction.

Altogether, it isn't, as I see it, a very cheerful human foreground. I am reminded of some of the pictures of Thomas Rowlandson. Look at the whole, from a distance, and you see a cheerful, almost idyllic composition of man blending with nature. Look a little closer and you see a tangle, a mass of men and women, a society, that is confused and sordid. Look still more closely, at an individual, and you may find something humanly appealing, comic, yet somehow heroic. I am coming to more individuals shortly.

And a past.

I used to smile at local history; it seemed appropriate only for country parsons and retired editors and professors. Here I have discovered my error. I have not yet had enough time and energy to learn more than what is common knowledge of our local history, but I have learned enough to know that its study need not be an escape; it can enrich and clarify the present.

Our county historical society is led by young people, and by a few older people who are notably young in spirit. They have put on historical exhibits, and a parade, that were both accurate and imaginative, and they are now building up a museum. My wife has been active, and our hired man and I have helped her to collect and return furniture. I must confess that this side of it leaves me rather cold; I am much more interested in migrations, buildings, old letters, and personalities.

The major trouble with local history is the difficulty in relating it to the history of the region, and of the country as a whole. Whenever I can see such connections, my interest is tripled. The doubtful fact that such-and-such a chair is "more than a hundred years old" doesn't interest me a whit, but if I can find out where it was made, and how it happened to get this far west, and no farther, I feel as though I were getting somewhere. There seem to be good general histories of the westward migrations, and plenty of local histories, if you can find them, but little in between. I can find out, if I want, what building stood on the northeast corner of Paint and Second Streets in 1838, say, but I shall find it extremely difficult to find out exactly what part this region played in the conquest of the West and in the growth of the nation. Another thing I want is an index to the newly-completed *Dictionary of American Biography*, made on geographical lines, so that I could read all the biographies of people who were active in this region.

The easiest and pleasantest way to study local history is to look up all the old buildings, and what is known about them. One wouldn't perhaps select southern Ohio as a field for the historical study of architecture, but one winter a professor gave a lecture here that was an eye-opener to me. Climate, tradition, migrations, economic elements, materials, local and imported workmen, pattern books and original ideas—all enter into the human stories waiting to be uncovered in hundreds of old private and public buildings within a couple of hundred miles of where I sit. Like most Easterners, I was provincial and smug. For all I know, there may be wealths of architecture, interesting historically and aesthetically, in such unlikely places as, say, South Dakota and Idaho.

Every time I go to our grocer to buy a loaf of bread, a pound of cheese, and some cans of dog food, that most humdrum of experiences is enlightened by the fact that I go to an old building on the site of a log cabin used by one of the early legislatures of the Northwest Territory. I can't drive up to our other farm at Bellbridge without seeing a burial mound made so long ago as to make Père Lachaise seem almost as new as a cemetery in the Bronx. When I feel a momentary nostalgia for France, or use a French word or two, and therefore seem affected, I like to remember that this region was first explored by Frenchmen, and that less than two hundred years ago it was the property of the King of France.

I like to imagine what this country looked like, then, and before, when it had not yet received all these blessings of civilization; or even in the time of my grandfather, who rode and drove horses, and who received and shipped much of his freight on canal-boats. I don't really like to, because I don't enjoy thinking of men as a vicious and depraved breed of animals, who can waste a countryside like a disease. I much prefer to look forward to a time when men can say, We nearly ruined this paradise, but now it is restored and protected; it is invaded only for good farms and gardens, for clean, sunny houses and factories that are appropriate, and for inconspicuous, efficient tools.

Only then, perhaps, shall we have the face, as well as the time and energy, to study local history.

An old house.

If you are one of our guests, you drive west from Chillicothe on U.S. Route 50, more pleasantly known here as the Cincinnati Pike. Two and a half miles out you find on your left a white house, red barn, and very neat garden owned by a man named Murgatroyd. There also is our R.F.D. mail-box, with Murgatroyd's and others. We turn up a narrow road to the left, past Murgatroyd's barn, and on up the hill. (In its memories, this road is as bloody as the sunken one at Waterloo.) It happens to be winter, with sodden brown fields, mists in the valley opening to the right, and purple-brown hills beyond. There are three new houses, disturbingly suburban at first appearance, but somehow, reassuringly, not quite that. Suddenly, at the brow, the hill flattens out into a plateau. Beyond a cattle guard and gate there is a pasture, a little too bare, with wisps of light-brown reindeer grass above the sod. There are a dozen huge gray trees, mostly white oaks and tulips, with a couple of great tree-corpses on the ground, only partly cleaned up. There are also a score of young trees, here and there, but they seem as pathetic as children among very old people. The road winds muddily towards a simple, rectangular stone house, with solid wooden shutters. The walls are pastel buff and gray, and the shutters a fresh cream color. In front of the house there are two large black walnuts, and around it other trees and bushes, including evergreens, but the house itself retains a certain sweet austerity. It is too simple to be pigeon-holed, but it suggests vaguely houses near Philadelphia, and Thomas Bulfinch. Out back, there are some neat gray sheds, and an old unpainted barn whose boards have turned silver.

Friends coming to see us are greeted by a salvo of barks from a collie and three black and white cocker spaniels that all come streaming around the corner of the house. But tails are soon wagging, and manners aren't too bad. These friends come in first to a large hall, running through the house, and glacial in winter. On the right there is the library, now unused, but pleasant enough in the summer. At least, there are flowers, then, the books have nearly all been read, and the pictures are often changed. Back of the library there is an old parlor, now unused except for potatoes, apples, odd pieces of furniture, picture frames, and the like. In the winter we live in the dining room on the left, and it looks it: a couch before the fire, desk, tables, chests, corner cupboard, a few curious pictures, bittersweet, a funny old hour-day-month clock on the mantel, knitting, books, papers, puppy toys, an odd glove on the floor. Beyond, there are pantries and kitchen. Upstairs, four bedrooms, a large hall, and a small back hall used as a bathroom. Everyone wonders about the net hung in the stairwell by my great-grandfather, to keep his children from breaking their necks.

Afterwards, we may wander out through the old garden with stone walks. There are the two small, fenced fields, now thin meadows, that we call the East Riding and the West Riding. Those woods on the western brow of the hill run down a steep cliff . Hidden on the hillside beyond the East Riding there is an old private graveyard, with a stone wall and a white gate. That lovely orchard to the south belongs to our neighbor, Ralph Stone. Those five hills beyond are in the Huntington hills, and they go all the way to Tennessee. We are on the very boundary between the corn and the corn liquor belts. It doesn't take much encouragement for me to show you our few cows and calves, sheep, ducks, pair of geese, and four new pigs. We got rid of the chickens on account of disease, but we shall start again in the spring. My wife says she will take them over. Now you can see the difference between this ram, here, and—but no; come on in and get warm. We may possibly have a spot of something to drink.

Now this place, Oak Hill, is not at all remarkable, in any way, and we are often puzzled and annoyed by the interest it arouses. Sometimes we feel like flat-footed museum attendants. Sometimes I fancy some of Peggy's relations and friends think she married me for a house, and got stung. An old college friend of mine came out here with me for a few days one summer years ago, and ever since he has been muttering about the place unintelligibly. I think it was an attack of symbolism. One young girl always shudders deliciously. "I always expect to stumble over a corpse somewhere," she says. "It's just perfect for a murder-mystery!" Some distant cousins always remark: "What a pity it is that you can't really afford to keep the place up." An old lady in town remarked kindly to my wife, one bitter winter's day: "It would be nice if you had a decent house to live in."

Now this old house, in its ninety-seven years, has seen a good deal of eating, drinking, sleeping, working, laughing, talking, listening, dancing, scheming, quarreling, lovemaking, child-bearing, teaching, sickening, dying, and weeping. It has also known emptiness, and silence, except for banging shutters, insects, and rats. Which is one reason I like best the remark of our friend Miss Mary Yates Bell, "It's a sweet old house, and I'm glad you're living in it."

And although I detest ancestor-worship, and various other forms of misplaced pride, and rhetoric, there is a poem by A. MacLeish that speaks to me. It is called "Men," and the last line is:

> We have lived a long time in this land and with honor.

With a few ghosts, mostly pleasant.

Not with great public honor, to be sure; merely, perhaps, without undue deference to the practical, and with a certain quiet, sinewy talent for graceful and sometimes happy survival.

My father's mother's father, George William Dun, was a Scot, born in Kilsyth, the descendant of clergymen, teachers, and merchants. As a young man, he emigrated to Philadelphia, imported dry goods, prospered, and decided to move west. He had brothers in this region and in Kentucky, and they all bought scattered coal and farm lands, and traded in cattle, horses, and mules. He made the move in 1838, and brought his growing family and his furniture across the mountains in carriage and wagons. His wife was named Louisa, "the daughter of a French gentleman named Duane." The house at Oak Hill was started in 1838, and finished in 1840. It was designed by the owner, with changes, in imitation of a house he had admired near Philadelphia. The stone, or at least most of it, was quarried here on the place. Like other early houses in this region, this one was placed on a hill because there was much malaria in the valleys. The place has sixty-three acres, and was chosen as a home; it has never been profitable as a farm. During the building of the house, the family lived in an older one, nearer town, called Dun Glen, which looks more recent because of some neo-Tudor details that were added later.

Sometimes I wonder about old George W. Dun. In his later pictures he looks strong, nervous, and melancholy. He disapproved strongly of at least one marriage in his family, though not of my grandmother's. He had a board walk built on the western edge of the hill, and walked there regularly, kicking off pebbles at one end, to keep count of the rounds. I think of him stalking there alone, wondering whether his daughters would marry poor fools and his sons would squander his property. In the forties he predicted without regret the imminent collapse of the Union.

In the fifties, my grandmother and one of her sisters visited relations in Scotland, and stayed there a year without once going into England. About 1860 she married a prosperous and respectable young man named David Smart, Scotch-Irish in origin, a wholesale grocer in business, a Democrat sympathetic with the South in politics, and an Episcopalian in religion. They lived in a gracious old house nearer town, and had two children, my father and my aunt. It seems to have been a very happy family life. My grandfather (I am proud to say) bought his way out of the Civil War draft, made occasional trips to New Orleans by steamboat to buy groceries, went into politics in a small way, and for eight years was Mayor of Chillicothe. (My friend Gerald Rowan's grandfather was his Chief of Police, and, as Gerald has pointed out, neither one of us has the character to be either a Mayor or a Chief of Police; we read books.)

When George W. Dun died, it soon appeared that his sons and daughters, for the most part, lacked his tenacity and competence. The coal lands, farms, and furniture were scattered and soon lost. I don't know much about these sons, but I have heard a few details. One of them collected naval prints (now lost), stuck his lapels full of pins, and had a passion for cleaning out culverts, which he would do for anyone who would let him, without charge. Another was a wag, full of schemes; without knowing how, he built a dry stone wall along our eastern border, and it soon fell down. Oak Hill itself was left to the oldest son, more competent than the others, who sold it to his mother, who left it to my grandmother. For about ten years, in the eighties, it was either uninhabited or rented to men working on the railroad. Briars grew up everywhere, trees fell down, and the house was infested with rats.

About 1884 my grandfather went bankrupt, and lost everything he had, including his home. His wife lost several farms, and all her property, in fact, except Oak Hill and a farm twenty miles away at Bellbridge. My father, who was at college, left without a degree and became a newspaper man. My aunt secretly studied typewriting (then a novelty) and got herself a job as a secretary. When she was about to begin work, she summoned courage to tell her father. He was offended to his depths, and told her that no daughter of his would ever hold a job. She accepted his will but I don't think she ever really forgave him. The farm at Bellbridge helped, and so did some rich cousins in the East and nearer home—God rest their souls. For the rest, my aunt squeezed pennies out of this place, and saved them: and no one is in a better position than myself to salute that achievement.

I often think of that "ruined old gentleman of the old school," in a white suit, with a little white beard, driving out into the country with his wife and daughter, to live the rest of their days in his wife's old home, the dependents of rich relations-in-law. The front pasture was overgrown, and cluttered with fallen trees and limbs; the old garden was in ruins; the back porch and the barn and outhouses were literally falling down; the shutters were peeling, rotting, and falling off; the grass in the yard was three feet high; and indoors, tenants and rats had left their filth. My aunt sat down on the front steps and wept.

But then she stood up and went to work, went to forty years of hard, grinding work. Almost single-handed, she cleaned everything up, and started a garden. She bought her first hens and cattle; she sold eggs, butter, and cream; she got up early and worked late, and saved every penny; she nursed her parents until they died, and then she lived on here alone, sanely and with good cheer, making her own life.

(My mother's people, who do not haunt this place, except in my unworthy person, were on the whole less conventional and more interesting.)

The presence of a lady.

My aunt Mary was a very small person, with a nice figure, a sensitive and gentle face, and gnarled and competent hands. She was reserved and unemotional, yet warm and generous, and deeply loyal to her own friends and own standards. One felt in her that sheltered and happy youth, that disaster, and that slow, solitary triumph over self and reality. In her there were blended extreme unworldliness and remarkable competence, the gentle reserve, innocence, and bewilderment of a lady of her time, and the stamina, fatalism, lack of squeamishness, and ingenuity of a good farmer. She would drive to town in her phaëton to make formal calls on her old friends, or on God, and then come home to wash milk pails, dust her roses, and treat a nasty wound on a horse. Some people thought she needed their protection and advice, but they never offered it twice. They had not seen her working with plants, animals, and men, and then cooking her supper, washing her dishes, and settling down, with her dog at her feet, to make up accounts, write letters, and read nursery catalogues or the *New York Times*. She used to say: "People ask me what I *do!*"

As a boy, I visited her here every summer, and for me, life began in June and ended in September. Luckily, she knew nothing about child psychology, and was both poor and busy. I trailed around after her and the hired man, and was allowed both to do any work I could and to amuse myself in any way that occurred to me and wasn't destructive. I rode for hours on end, and was taught and expected to take good care of the horse. I used to do errands, even, and go to children's parties, on horseback. I was taken calling, and to church, but at no other time did I have to dress up. The only times I was disciplined, and that sharply, was when I didn't show respect and courtesy to my elders and to workmen. In general, it was a matter of my feeling that I had a mind, feelings, and rights, like a grownup, and that my aunt found me pleasant and useful to have around.

After I went to college and to work I didn't get out here so often, but when I did, we had tremendously good times together. I knew that my aunt did not approve of many of my ideas, or of much of my writing, or of some of my behavior elsewhere, but it was always quite as clear that she thought that was all my own business, which she did not feel competent to judge. She respected me as I respected her, and though it was never mentioned, and shown but timidly, the old affection had only deepened.

The last time I visited her, I arrived, as usual, about six in the morning, and came out in a cab. When my aunt came to the front door I lifted her off her feet. Afterwards, an old colored woman in the kitchen told me that my aunt had hurried into the kitchen, ashamed of her wet eyes, and saying half to herself, as though in excuse: "No one has ever done that to me before. No one has ever lifted me off my feet before." One high moment of that last time was when we went to an exhibit of the Garden Club. My aunt won a couple of ribbons for some huge, pale zinnias, and lost another ribbon because she insisted on using a Directoire vase I had sent her from Paris. We had many a good laugh over that, and over other little things. She seemed well, but very tired.

Two evenings after I had left, she had her solitary supper as usual, washed the dishes, and sat down to read a paper. Her only help, a hired man, had gone away three hours before. She felt tired, but happy, because that afternoon she had got a telegram saying that her niece's husband, of whom she was very fond, had come through an operation very well. At eight o'clock she telephoned her doctor and said she had a strange and frightening sensation of choking. Would he please come out quickly? The doctor's own hand was in a bandage, so he got another physician, who had also treated my aunt for many years, to drive out with him. They arrived in about five minutes.

Jack, my aunt's collie, is ordinarily very gentle, and he knew both of these men well, but when they arrived, he stood just inside the screen door, barking, so savagely that for a few minutes they dared not enter. There was no other sound, and only one lamp burning, on a table in the hall. Finally they calmed the dog, and got in. Still growling and trembling, he ran into the dining room and hid beneath a couch where lay my aunt, alone, as she had lived. As everyone lives. Everything was in immaculate order.

Now I am not at all psychic, but sometimes I long so keenly for a word with my aunt that I imagine her, feel her, suddenly appearing at my side, in the barn, or in the garden, or in the pasture. It does not seem ghostly, or weird. There she is, in her broken old shoes, her clean dress, her big straw hat. We look at each other, and for a moment say nothing. Then we rush about the place, both talking at once. It is all very practical, and about things that don't make good dinner-table conversation. I show her the new things, and try to reassure her about money, and the future. We laugh a good deal, as we always did. And here comes Peggy! There is a light in my aunt's eyes, and she shakes her fist in that odd little gesture of emotion and delight, and is gone.

Once when my own small nephew was here, he cut down a weeping willow tree that had been given us, and that we had set out with care, and nursed for months. I found it, a whittled stick, on the floor of his room, and I could have tanned his bottom until he screamed. As I stood there, looking at that stick, I felt my aunt beside me, smiling at me. I remembered a certain hen, and a beebee gun, and a hatchet that another small nephew had played with twenty-five years before.

Another kind of inheritance.

But there is another kind of inheritance, less important, which troubles me.

After my aunt's death, by the grim irrelevance of a will, my sister and I became owners of Oak Hill (one-quarter and three-quarters respectively), of the farm at Bellbridge (equal, undivided interests), of some livestock and furniture, and of a few thousand dollars apiece. Also, my aunt left me her dog, Jack, with money especially set aside for his support. It was more complicated than this, because two other wills and a generous gift were involved, but this was the final result.

When romantic story-tellers want to enrich suddenly one of their characters, they make the deceased a distant cousin in Rhodesia or somewhere, who had plenty to spare, and whom the legatee had never seen. The recipient never has any doubt whatever about the institutions of private property and inheritance. In all this, the story-tellers show their usual intuitive wisdom.

Oak Hill may be called personal property, in that it is practically unproductive, and, God willing, will not be sold. About it I have no qualms, because I think strictly personal ownership should be based on personal interest, of which we had as much as anyone. The farm at Bellbridge, on the other hand, is a part, however small, of the land that feeds the American people. I think it should belong to the people as a whole, or lacking that, to the man who has farmed it all his life, on shares, and whose father farmed it before him. I feel guilty every time I get a small check from that farm. It has been owned by my people for five generations, and not one of us has ever done a lick of work on it. I retain a basic prejudice, not acquired from books, that work is more important than capital risk, and the good of the people as a whole, as producers and consumers, more important than either. All I can do is act as responsibly as possible towards the farmer, the land, the government, and my generous but absentee partner—and more or less in that order of importance. Much worse, from my point of view, are the few thousand dollars, which make me one of the shareholders in whose interest employers exploit labor, which are active in a market directly opposed to my interests as a producer and consumer, and which have "made" me much more money than I have earned as a farmer or writer. I have spent some capital, and shall soon put the remainder into land or other means of production that I can work myself, or into government bonds.

If any radical or conservative thinks this course of action is hypocritical, I should like to ask him what he would have done in my place.

I remember discussing inheritance, several years ago, with a wise and humane French banker, who explained to me the

French idea of the relation of property to family, and the superiority, in the public mind, of inherited property to earned. In this country and England, conventional ideas on these matters are similar, if less realistic. I respect French civilization profoundly, but I hope that in time we shall attain something even better, something in which only personal property is inherited, something in which the naked human being is superior to all property, and incidentally, something in which there is no such thing as an impoverished "country gentleman."

Above all—and this is much more idealistic, and goes much deeper still—I hope the time will come when all young people can grow so freely that their chances in life cannot be altered by the survival or death of older people whom they love.

II. Ally

Two frightened marionettes.

A life of solitude in the country is by no means to be feared, even by a woman. If she has physical strength, and some courage and ingenuity, a woman can live alone on a farm and operate it herself, with such hired help as she can afford. During the last seventeen years of her life, my aunt lived alone in this house, without plumbing, electricity, or central heating. She had one hired man, but at night she was alone. For a while she drove a horse to town, but later she was limited to infrequent taxis. She never kept a gun, and she said that if she were robbed at night she would merely give the burglar anything he wanted. She was active and objective in temperament, and maintained a standard of sanity and happiness not often reached by maiden ladies living with other people in cities, in much greater comfort. Nor was her case an isolated one. For years one of the best dairies in this county was owned and operated, and much of its actual work was done, by a solitary woman.

A man, of course, need be even less deterred. Our tenant at the other farm is a bachelor, and lives with a bachelor brother. They don't milk their cows, so their kitchen work is lessened, and they have some regular help from their sisters in town, but they do most of their housework themselves, and their home would put many a housewife to shame. There is a story by Dorothy Canfield Fisher, in *Raw Material*, that I find very moving, about an old man who lived and died alone on his farm in Vermont. One week I lived here quite alone, taking all the care of animals, garden, house, place, and myself. This is nothing to boast about, but I found it exhilarating to discover that I could certainly do it. There is a widower I know who has done all this, and much more, for years.

Of course there are other possibilities. Real farmers, of course, never have the money or taste for conventional, parasitic mistresses. The urban possibility of living more or less privately with a real friend, without the complications and responsibilities of marriage, is ruled out by small-town morality, and by party-lines. Country people don't judge and condemn as townsfolk do; their curiosity is friendly, but it is naturally very active. A lover in the country might be like a wife or husband there, although ostracized in town.

In my case, I lived here for six months, as a bachelor, with my aunt's Negro hired man, his wife, and their four children, and I was far from miserable. I have a very keen taste for solitude, I was unconscious of the crudities of the housekeeping, my social life was almost nonexistent, and I was hard at work at this typewriter (on another job) and in the fields. I worked all day, read every evening, and in bed was generally too tired to feel the solitude too painfully. If I had carried on indefinitely in this way, I should probably have been fairly happy, and unconscious of my drift to emotional drought and neuroses. Either that or I should have emulated the monk of Siberia.

As it was, not long before leaving the East I had had an extremely good time, and in a joking and sardonic way had fallen in love with a fairly old friend. That summer, I invited her out here during a visit of my mother, and for her sins she came. Fear, intellectual conviction, and even good sense were all overcome by an inner compulsion (abetted by circumstances) that drove us both through painful and futile inner conflict to a scene that took place in a typically New England drawing room one cold, gray, sleeting afternoon the following winter.

The room was decorated with flowers, and was full of people in their best bibs and tuckers. Two little girls were inspired by the church-like air of the assembly to find a hat and take up a collection for themselves. There was no coffin, but in front of the fireplace stood a blue-eyed, white-haired gentleman in an academic robe, and facing him stood a young woman in white and a young man in black with a face temporarily the color of good old Roquefort cheese.

The night before, we had been foolish enough to count up the hours we had spent in each other's company, both alone and with other people, up to that time, and the figures had not helped our already somewhat shattered nerves. I think we reached a grand total of something less than a fortnight. We had known each other for nearly five years, but we were still comparative strangers. We had been in love when we had agreed that we had to submit to this joint fate, but social conventions and mechanics had dissolved so much of an emotion that we had both felt so often before, for other people, that little of its power to impress, or impel to action, remained. We were both more than thirty years old, and had faced the prospect of going on to our graves in a more or less celibate manner much more cheerfully than we faced the prospect that now confronted us. Neither of us thought that any man and woman could reasonably marry until they had lived with each other for a couple of years, at least. I had a horror of being sold out to a wife, that is, of mortgaging my deepest feelings, instincts, and ideas. I loathed every social aspect of an engagement and marriage. I felt morally and economically unable to afford children, and distrusted Peggy's agreement on this point. Peggy had had a good job that she had liked, and was giving up that, an admiring circle of family, friends, and acquaintances, sailing, and all the social and cultural amenities of independent life on a good salary in the East, for a relatively impoverished existence, with a strange man, on a farm in the Middle West. I was hardly beginning to get used to farming and country life, and was anything but eager for the additional, superlatively complicating element of a bride. Various men and women I had known had lived alone in the country quite happily, and I thought I could. I still felt a good deal of weight in the old argument that anyone with artistic aspirations and little money could not well afford to marry. And so on, almost indefinitely.

In the face of all this, we were brought to that pretty pass, in that drawing room that winter afternoon, by a compulsion, within each of us, so deep and so strong that we had recognized our helplessness before it from the moment of its first appearance. I could only describe this compulsion in terms of anesthesia or hypnosis. It certainly had nothing to do with freedom of choice. If we had not felt it, we certainly should not have married. We feel fairly confident that it was not merely one of the many elaborate and subtle tricks of sex. My economic circumstances, over which I had had no control whatever, made marriage just barely possible, and we should not have married had I been any poorer, but we certainly were not rich enough to marry just for a new sensation. We trusted each other, and I at least found that a fairly new experience, and one without which we should not have married, but it certainly was not one that, alone, could have driven us to such a desperate measure.

It was the best step I have ever taken, or been forced to take. This is the sort of thing, I imagine, that makes people begin to mutter mystically about the will of God, without always remembering the superlative difficulty of discerning it.

Over the doorsill, on foot.

Getting acquainted with your wife or husband, when she or he has been drawn for you from a deck of several thousand cards, and you have submitted to her or to him as one submits to an anesthetic, is a little difficult under any circumstances. Those that we encountered made the process quicker, but not easier.

We decided that since we were about to retire to the wilds of Ohio for an indefinite period, we'd not go into the conventional solitude, but spend a week or ten days in New York, eating, drinking, and bathing well, and going to as many plays, and seeing as many of our friends, as possible. If morality consists in deferring pleasure, we could not have done anything more horribly moral. Our rather stupid error consisted in forgetting that our bodies and nerves had already been reduced to quaking shreds, so that food, drink, and drama didn't mean much, and in ignoring the fact that we had no single friend in common, and couldn't very well become friends of each other's friends under such odd circumstances. The bride or bridegroom was, in company, always a painful curiosity, both bored and irrelevant. It became so monstrous as to be almost funny, even to ourselves. But oddly enough, our damnable virtue paid. Later, we had a lot to talk about, and relish, and second meetings with each other's friends were much easier. I imagine most weddings and honeymoons, if people would admit it, are both very painful at the time and delightfully comic and memorable afterwards.

When we got "home," to Oak Hill, the countryside was brown and dreary, with dirty remnants of snow and a heavy winter fog. I was so excited when we crossed the county line that I could hardly contain myself, although I knew that for Peggy this "home-coming" must be something very different. She had visited me only that once, for a few fatal days, in the warm green glory and peace of the summer before. We bought a last bouquet, and then began to keep accounts. When we reached the house, I forgot to lift my bride across the doorsill. I did not really regret it, because we both knew that she would always have to walk on her own two feet.

The "servants" were almost strangers to Peggy, and soon enough hostile to her, and the house was as cold and damp as a tomb. Poor Peggy! She caught a cold, which was followed rapidly by intestinal grippe and asthma. She retired to bed, where she was more or less cared for by a strange husband, strange Negroes, and a strange doctor. There was no plumbing, no heating except by grate fires, and no electricity. There were filth in the pantry and kitchen, rats in the wainscoting, and icicles in the privy. The water in the cisterns was low, gray with soot from the snow on the roof, and of doubtful sterility. At this point, one cistern caved in. Things like this went on for months, until spring, which, like the summer that followed, was so soaked with rain that we found ourselves living in a marsh on top of a hill. As soon as they had been hoed down, the weeds in the garden stood straight up on their feet again. Add the brief but passionate jealousy of my dog, the sudden, violent death of Peggy's first dog, the deaths of scores of chicks, a "holiday" in Chicago during which everything at home went to hell, the necessity of being correct in a painfully sensitive though warm provincial society, Peggy's flight home to Mother and Ocean for a month, the sudden, painful departure, at that inopportune time, of the servants, and a number of other little experiences of the kind, and you get what might possibly be called an Introduction of Bride and Bridegroom.

When Peggy returned that autumn, with her father (the gentleman with blue eyes and white hair, already mentioned), we might have said to him: "Now, Sir, if you please, you may marry us. We think we have some small right to it, and we think we'd enjoy it."

Poem deleted.

Peggy was born in New England, and later returned to it, but luckily for her, as well as for me, she spent her childhood and early youth in Baltimore, where there are a lot of Catholics, Negroes, Southerners, and other people who know and care very little about the rather peculiar virtues and failings of the direct descendants of the Pilgrims, and where there are also the soft, luxuriant valleys and blue hills of Maryland. The early married life of this clergyman and his wife, and the childhood of their four offspring, seem to have been exceptionally fortunate and happy. There were money and security, but not too much of either. There were good friends, good work, and good holidays on Cape Cod and in the recesses of Virginia and Maryland. There were those fabulous journeys, with piles of babies, children, baggage, and dogs. There were the long evenings, spring and autumn, with a swarming and happy life centered on one of those immaculate white front stoops of Baltimore.

Then the return to New England, with all the agonies of adolescence in grim towns where it seemed to take people years to consider themselves acquainted. Then college, and being kicked out, simply because she had always been too sheltered and too quick-witted to be forced into the difficult activity of thought. Trips into the South and West, the West Indies, and Europe. Finally a job, and an apartment of her own in Boston.

From my point of view, this job was amusing and important. It was with the Girl Scouts. Peggy seems to have liked it, and to have been good at it. There is nothing like a job for learning what a dollar means, and how to get along with, and finally perhaps to appreciate, disagreeable people. Peggy also picked up some practical and theoretical knowledge about such matters as child psychology, amateur theatricals, and the care of latrines. The ingrained consciousness of hygiene sometimes drives me to a frenzy, but it may save the lives of our guests. Luckily, the girlish idealism and ruthless efficiency of Girl Scouting were tempered by other experiences. Still, one of her aunts and I have formed a "League for Sitting Down at Will, despite Peggy's Hyperthyroid Glands." (Final notes for young men thinking of marrying Girl Scout executives: These girls in uniform retain the conventional feminine attitude towards rats, mice, wasps, hornets, spiders, and other matters. In case she carries on, as an unpaid worker, the outskirts of meetings provide much good comedy. If she is good at it, and you insist on writing books, or something, or die of lockjaw, she can get a paid job again.)

All these words don't seem to mean much, or if they do mean anything, they miss the essence. I should like very much to picture much more fully and directly a woman whom I find as interesting as delightful, and who is really the heroine and central figure of this whole picture, this whole book. But after months and months of effort I don't seem to be able to do anything better than this. When I was in college I could write really rather nice poems about dream women, but the more women I knew, and the more I loved them, the less poetry I could write about them, and here I am, writing about everything else, including the Girl Scouts. I wrote a prose portrait of Peggy that I thought was very good: detached, honest, "sympathetic," well phrased, full of life, and almost as amusing, nice, and exciting as the original. I have been assured by one of the best authorities I know—not Peggy—that it is "really atrocious, in the most egregious taste." I thought it was lyrical, almost youthful, and another good reader assured me it was the most cynical and brutal portrait of a woman by her husband, with whom she was still living, he could imagine.

I shall have to leave it out, after all. I can only hope that in the rest of this book, where she appears in passing, obliquely, doing this and that, there will survive some of her . . . some of her . . . damn it all, some of her beauty. If none of it appears in these pages, this book is a dud, and I am . . . well, I am damned.

And as for you, Peggy, you really ought to have married a poet,

> Telling you men shall remember your name as long
> As lips move or breath is spent or the iron of English
> Rings from a tongue.*

---

*A. MacLeish: "Not Marble nor the Gilded Monuments."

III. Principles

An old field of learning, revisited.

Sometimes our guests have asked us: "But how did you learn so much so quickly?" This is an embarrassing question that we should not care to have asked in the presence of our neighbors, because of course we have learned very little indeed. It takes about thirty years of active work, experiment, observation, and study to make a passable farmer. Meanwhile, we try everything that seems reasonable, watch the costs and results as carefully as we can, pester the lives out of our neighbors for information and advice, drive all over the countryside with our eyes open, refer major questions to our very intelligent and helpful County Agricultural Agent and to the State University, and read books, magazines, and the invaluable free pamphlets of the United States Department of Agriculture. We have been doing this for three years, but even so, hardly a week passes without our being confronted with some new problem or other that makes us look and feel as ignorant and helpless as though we had just arrived from Manhattan.

One friend has written to ask us about the very first impressions, the most elementary questions and situations that confront a beginner. This question is answered, in part at least, elsewhere in these pages. I shan't soon forget my first morning alone as a farmer. I was a schoolmaster, out here on spring vacation, trying to make up my mind about whether to try just this one more career. I had driven out from New York in an old Ford, and was exhausted. About six o'clock that first morning the hired man woke me up and told me that Elaine, one of the cows, was badly off her feed; he wanted to know what to do about it. I got up, assumed a knowing air, looked at the cow, found out what the hired man knew, consulted a neighbor, went to town for a bale of the best alfalfa, and with this coaxed her back to a less expensive diet. It is always like this. Life on a farm is always going on, and questions are always arising for solution, questions of diet, disease, breeding, birth, planting, cultivation, harvesting, sale, repairs, reconstruction, drought, flood, wind, hygiene, mechanics, and all the rest. In time, of course, one learns to look ahead a little, and to solve a few problems before they arise. The quality of a farm is measured accurately and beyond appeal by the quality of the farmer, that is, by how much he is growing, and enjoying life, himself, and not by his efficiency. However, his own growth, health, and pleasure are determined in part—though not wholly and always, by any means—by the amount and quality of his anticipation.

The application of knowledge is the thing, of course, and city-bred farmers like myself, with an academic background, are apt to be seduced by the simple delight of learning. Often enough I have been reading about fruit flies, or artificial insemination, or some damned thing that had little or nothing to do with my farm at the time, when my own chickens or garden needed attention. However, I am not wholly ashamed of this, because my neighbors err in the other direction, and because if my farming is a mess, I always find it a most entertaining one.

When I went east to get married, I went on a bus, and I

got to talking with a movie publicity man. I was all dressed up, and looking out the windows, while he talked candidly about farmers.

"What I can't get," he said, "is what these hicks do with themselves. They can't work all the time, and they can't likker up, and —, all the time, and they ain't got no money. What the hell do they *do?* Jeez, they murder me!"

The answer is that even the poorest and most ignorant of them probably get more fun out of their work, itself, than most city workers get out of theirs.

And once a very wealthy "farmer" told me how foxhunting relieved, for him, the "dullness of country life."

Boredom is a mysterious matter, more appropriately handled by physicians and psychologists than by me. In myself, I find it caused less by the immediate situation, less even by more remote and less obvious circumstances, than by some obscure inner weakness, fatigue, and dullness in myself, or simply by something in the air pressure and weather. However, it seems to me that it would take the most extreme and outrageous poverty, or wealth, or weakness, or lack of curiosity, or ineptitude for reading, watching, and conversation, to make country life dull. There is too much to learn.

The variety of knowledge demanded, if you are going to stay off relief, is astonishing. You may not know what the words mean, and you may never look into a book, but you have to have a working knowledge of some of the principles of genetics, zoology, botany, morphology, agronomy, chemistry, physiology, embryology, pathology, evolution, hygiene, carpentry, mechanics, physics, bookkeeping, accounting, economics, and so on.

Even a good farmer's knowledge of these subjects is very limited and empirical, but as far as it goes, it is not apt to be superficial. It is easier to learn enough about something to talk about it entertainingly on a typewriter, or in a classroom, say, than to learn enough to turn that knowledge into money by way of the soil and animals. A farmer may use his empirical knowledge most, but his intuitions, based on experience unconscious or forgotten, and aroused by sympathy, are to me more exciting. They may even be more important. I like to see a good hand at a job, but I like even better to see a farmer at work who is so close to his animals that he knows what they need without any kind of thought at all.

In all this, Peggy is much more alert and less absent-minded than I am, and has saved plants and animals more than once, when they would have died, for all of me. But sometimes, when —as at lambing time—I spend a lot of time with the animals, and brood over them, I can begin to feel, a little, what is going on inside them, and if there is anything more exciting than this feeling into an alien body and mind, I don't know what it is. And it is open to everyone.

In praising and aspiring to the knowledge of good farmers I do not mean to veil and condone ignorance. I have learned more from my neighbors than they will ever learn from me, but whenever they hint that science has nothing for them, or that nothing new to them can be any good, or that knowledge other than practical and agricultural is foolish, my sympathy and admiration cool perceptibly. If there is anything that keeps me from an abject inferiority complex, in relation to more experienced farmers, it is the idea that my greater respect for science, and ability to hunt out humbly (if not always to use) pertinent information, may compensate, in part, for my relative weakness and inexperience.

Another idea that helps my ego is that because of all my apparently irrelevant knowledge and experience, I may actually be enjoying farming more than most farmers. I may be weaker at the start, which is good farming, but I may be stronger at the finish, which is enjoying life. For instance, I like to look at landscapes, thinking of paintings I have seen or may see. If more farmers could do this, they would spend the old A.A.A. 20 per cent of their time simply staring, and production would decrease accordingly. I offer this idea, modestly, to the government. It would also help our farmers to become what—if they only had a little more regimentation and the sweet craziness in the head—they might be, namely, the happiest men on earth.

I must point out that learning in the country is immeasurably slower than in the city. In my first three weeks in the publishing business, and my first three weeks as a teacher, I learned more about publishing and teaching than I have learned about farming in my first three years on a farm. The reasons for this are that the life processes, which only expose error, are so slow, that income has so slight a relation to skill, and that isolation keeps one from being checked up on by colleagues and bosses. In about thirty years it no longer embarrasses you acutely to claim that you are a farmer. And then where are you? You are crippled with rheumatism and ruptures, the bank owns your farm, and your children want to go to medical school.

The hunting cap vs. *the derby hat.*

Before I talk briefly about the management, principles, processes, and business of farming, I must emphasize a few of my own disqualifications and prejudices, so that the reader may be able to discount, as he sees fit, what I have to say.

Not only have I been farming only three years; my farming is extremely small potatoes. I make the decisions and do a good deal of the work on this farm, Oak Hill, but it has only 63 acres of rather poor land, mostly in pasture and woods; our other farm, at Bellbridge, has about 184 acres of much better land, but it is farmed by a very superior man named Kincaid, to whom I am only a legal and financial partner and consultant, as well as a friend. Furthermore, you must remember, my political and economic notions are definitely radical. Finally, I have small talent or taste for management, and am much more interested in production and use than in profit, money, and trade. This last prejudice is rather strong, because it has at least three sources: a fellow feeling for producers, manual and mental; vestiges of what might be called the old, aristocratic distaste for salesmen and entrepreneurs; and economic ideas.

The reader may share some of these prejudices; if so, he will have to be doubly careful to discount, and to look at the other side. Indeed, I think it likely that the reader will have the same prejudices, because I can't imagine that anyone with ambition of the ordinary kind, and with talent and taste for management and business, would be so foolish as to turn towards farming, and, incidentally, towards this book. The only people I know who make what educated and lucky city people would call a decent living have farms of a thousand acres or more. My kind of farming is not for incompetents and weaklings; it is for people who have no interest in "getting ahead," who like animals and plants more than machinery, processes more than figures, solitude more than most company, and a hunting cap more than a derby hat.

Not that we can escape decisions and management: far from it. We find ourselves involved in intricacies of management that, however petty in scale, are not fundamentally unlike those that confront Executive Committees and Boards of Directors. We can wear dungarees, and go without neckties, and do our figuring outdoors, on the backs of envelopes and weight slips, but we too have to scheme, calculate, and manage. We too have to think of source materials, processes, labor, outlets, accounts, plant and equipment, taxes, the law, and all the rest. We are peculiarly at the mercy of markets, because our methods and plans cannot be changed quickly, and because we have few sources of business information. Also, our opponent, nature, can make our best strength and brains look very small and futile, overnight. However, she is also our partner, and she has incomparably greater dignity than any other opponents and partners. Besides, even Thoreau had to scheme and sell, and to decide whether he was going to pay his taxes or go to jail.

In my experience, the best way to get around this whole difficulty is to play the manager, employer, partner, salesman, trader, business man, taxpayer, and responsible citizen as one plays a game, or an unsympathetic rôle on the stage. One can study it objectively, and play it hard, and shrewdly, even with a certain flourish and pleasure; and all the time, without for a moment betraying or damaging the performance, one can hint, can make it quite clear to the intelligent and sympathetic, that it is all a game, a part to play. I find that tradesmen in town, and others, are apt to take all this with discouraging seriousness, but that some of my farmer neighbors, in seriously giving me advice, or in shrewdly bargaining with me, have the gleam in the eye that makes all the difference. Not long ago, in this manner, one of them actually asked *my* advice! They don't like these parts any more than I do; they know that all the world's a stage, a carnival, that even deeds, debts, mortgages, and taxes are tawdry stage properties, and that "We are such stuff As dreams are made on; and our little life Is rounded with a sleep."

A few principles from life, and from a poem.

In thirty years I shall know, I hope, much more about farming, but in three I have acquired a few basic ideas that those even less experienced may find useful for consideration, adaptation, and perhaps rejection. I have acquired them in the ways I have stated, and—not so oddly as it may seem—from a poem called "Build Soil—A Political Pastoral," written by a farmer in Vermont. Our farming is different from that in Vermont, or Cape Cod, or Wyoming, or California. On the wall in our County Agent's office there is a large map, in colors, showing roughly the kinds of farming done in the different areas all over the country. As I remember it, this region is colored as one of mixed or general farming. In any case, I think we may have a good deal in common with farmers in Vermont or California, now, or even with farmers in Greece a couple of thousand years ago.

The first idea is fairly simple, and all the others, with exceptions and qualifications, can readily be inferred from it. It is this: Under present circumstances, every farm should be as self-sufficient as reasonably possible.

This implies, first, that the land should be conserved. In one way or another, nearly everything that is taken away must be put back. Otherwise, your farm will run down, more rapidly than you think, and then where are you? On relief.

The second major corollary is: Buy as little as possible, of fertilizer, tools, labor, food, clothing, health, pleasure, and everything else. Every penny you spend, you have to earn by growing or making and selling something, and on a farm, using is much easier and more natural than selling; also, much safer.

The third is: Sell as little as possible; push the product along towards use as far as you reasonably can, on your own place. This makes the best use of waste for conservation, of by-products for feed, and of equipment and labor all the year round. It saves from others a few of the profits of transportation, speculation, and processing.

The fourth is less obvious, and more debatable: Take part in Coöperative buying and selling. This is the method of making up for the smallness and isolation of the economic unit, and of extending the emphasis on production and use, as opposed to buying, selling, and profit, beyond the line-fences of the farm.

Before I go on to applications and misapplications, failures and successes, in my own experience, I must make a very few qualifying and explanatory remarks.

I do not advocate this policy of isolation and self-sufficiency under all circumstances, and in all times and places. On the contrary, I think that the recent history of agriculture in this and in other countries indicates a strong and necessary drift towards larger and larger units, towards specialization integration and planning. Some farms, regions, machines, and people can obviously do some things better than others. I should like to see this whole county, region, nation, and world become one large farm, democratically controlled. However, I do not believe in the efficiency, in this direction, of the market, and of trade, which never has been, and never will be, free. I especially do not believe in their efficacy for the individual. Dependence on the market means dependence on loans, mortgages, banks, and charity, and with international trade what it has been, and will be, there are going to be no Rockefellers of agriculture.

Neither do I advocate self-sufficiency to a fanatical extreme. Since other people can make and do a great many things better and more cheaply than I can, I prefer to buy their goods and services, and pay for them, if I can, by specializing myself. Once I was taken to call on a man who had a farm on which he hoped eventually to be quite self-sufficient. He was not only butchering and canning; he was carding, spinning, weaving, sewing, and making tools, soap, and everything else that he could. The investment in equipment was colossal, he didn't have time enough for any one thing, and he wasn't young. He seemed to me timid rather than shrewd; he had reduced the whole idea to absurdity.

The last idea, on Coöperative buying and selling, is open

to dispute in which I should delight to engage, but I don't think this is quite the place for that. Later in these pages I may be able to tell briefly of my own experience with Coöperatives.

A few attempted applications, in general.

Whatever his ideas of this kind, conscious or merely implied in action, a farmer's work and life are determined by the quantity, quality, and nature of his land and equipment, by the climate, by the markets, and by his own age, strength, knowledge, experience, and personal predilections.

Oak Hill has a layer of sour and heavy clay, about six feet thick. Beneath this there is a layer of a sandstone called Berea Grit, from twenty to thirty feet thick, with unreliable veins of water. Beneath this there is dry shale. The top of the plateau is slightly concave, and the drainage is very bad. Our outbuildings were in poor condition, and these we have taken the capital to replace and repair. We have a moderately good collection of garden and carpenter tools, and a couple of decrepit plows, but no other farm machines, and no team. I have already given some idea of the climate, in its reported norm, and in its more familiar extremes. For markets, large-scale farmers can buy cattle in Kansas City and Chicago, and ship to the larger markets in carload lots. The rest of us are dependent on local weekly auctions, but here we are at no relative disadvantage. We have to sell our smaller quantities of grain to local elevators, but here too I imagine we do as well as the others. Livestock and grain prices are notoriously flighty and unpredictable, and some better knowledge in this field can be worth more in money than intelligence and industry in producing the goods. Our prices are very sensitive to Chicago and world markets; but the spread is variable and considerable. We can do some Coöperative buying, as of feeds, fence motor oil, etc., and one kind of Coöperative selling, of wool. Fruit and garden produce can be sold to advantage only by trucking considerable distances to larger markets. Other markets, for specialties of all kinds, depend on the ingenuity of the producer, which has little play against chain distribution, in a community largely agricultural and now crawling but temporarily from a fifteen-year depression. Peggy and I are in our thirties, not towers of strength, and relatively ignorant and inexperienced. We prefer working with animals to working with crops. Such, in general, are the materials and conditions.

Our fields are not large or good enough to attempt the familiar corn-wheat-hay rotation of the bottoms. Tiling and liming would be good, in any case, but we have not felt sure that the expense would be repaid. An orchard or vineyard would be feasible, but the requisite knowledge and capital for either are well beyond us. This practically reduces the possibilities to grazing and specialties. A dairy would be possible, and even attractive, but the capital expense would be something, and we still doubt whether we should have the necessary moral and physical stamina. Grazing means either buying new stock every spring, gambling on plenty of rain and rising prices, and selling out entirely in the fall, or else raising as much corn and hay as possible, buying the rest, and expanding in the spring and retrenching in the fall only slightly, according to the prospects. Chiefly because we like breeding and raising animals, we have chosen the latter course. We have had from six to a dozen head of cattle, including two or three cows being milked; we intend to cut down to two cows, and keep the calves not much more than a year. We bought seventeen sheep; now we have thirty-six (ewes, ewe-lambs, and buck), and shall let them increase, perhaps, to forty or fifty. We have garden, chickens, and ducks for our own use, four pigs to eat garbage and skim milk, and four dogs for companions.

On conservation, we have done nothing to brag about, but we have done something. Our woods need cleaning up and thinning, but since they are on a cliff, this will be extremely difficult. Meanwhile, they are not pastured, and are replacing themselves. The front pasture, after use for nearly a hundred years, has lost three-quarters of its primeval trees. My aunt began replacements some years ago, and we have carried on, less because we think of "generations yet unborn," which we do not, than because we simply enjoy planting trees. If I ever decide that I can afford to give up that much pasture, I think I can get the C.C.C. to fence off a few acres on the east hillside, and set them out with very young pines, ostensibly for erosion control, although the need for that, there, is not yet serious. In the two years that I grew corn in the fields, I put in some fertilizer, and since the corn was first drowned, and then burned up, some of the fertilizer is presumably still there. So far, we have used so much manure on the garden that the fields have not got any, but their turn will come. Since we have to buy some hay and fodder, there is, in what comes from the earth and goes to it, what might be called a favorable balance of trade. Last summer, in the drought, I over-pastured slightly, but this was practically unavoidable, and better farmers than I am did the same.

The second corollary, against buying, we have violated right and left, but on the whole quite consciously. Our personal expenses, though still large, have been steadily reduced. There remain a number of practically fixed charges; these include the upkeep of a big old house with outbuildings and waterworks, life insurance, taxes, medical charges, automobile expenses, and so on. One of our largest expenses is in hiring labor. All year we have a hired man, now getting eleven dollars a week. In the winter, he works half indoors and half out. One summer, with guests and illness, we also had a girl in the kitchen at four dollars a week. We could do all the work ourselves, and we may come to that, but most of the specialties possible would require some paid help. In the matter of feeding as completely as possible from the place itself, we are more successful. A good farm is a complicated network of alimentary canals between the earth and the earth, operating in a manner that would have delighted Hamlet, or would be suggestive to a surrealist painter. Our system requires the purchase of some hay, fodder, chick feed, meat, groceries, and dog food, but not as much as you'd think, considering the hundred mouths we feed, and there is little waste or leakage. The livestock all feed the earth, and so shall we have done, when we move the privy, and plant an apple tree on its site. The livestock supply us with milk, cream, butter, and a little money for groceries. They also feed the dogs, fowl, and pigs with skim milk and undigested corn, while we feed the fowl and pigs with garbage. The complications are endless, and practically everything but tin cans is turned into manure or money. I could elaborate this physiological nightmare by introducing the thefts of the animals from each other, insects, bacteria, and death, but I shall refrain.

We have sold cattle, old ewes, wether lambs, chickens, wool, a few eggs, a little cream, instruction in French, and words on paper. We have bought dressed meats, but not enough, so far, to make butchering feasible. The steers are sometimes only grass-fed, that is, not finished off with corn for butchering—but this violates the third corollary only when enough corn for finishing is grown on the place. You can buy corn, and make a specialty of feeding, but so far I have not considered any more of this than we do now, economic. (Right now, April, 1937, corn is worth $1.25 a bushel, and top beef, on the hoof, about $10.50 a hundred.) We have bought some woolen goods, though not much. Peggy has a brother-in-law, a worsted manufacturer, who offered to trade us worsteds for fleeces, but we thought we could use the money better. The rest may be called, with some generosity, finished products that we can't use ourselves. My French pupil flunked his examination, but if the instruction wasn't finished, the pupil just about was. The writing was not as finished a product as it might have been, and I spend too much money on other people's words, but then, I do not, like Olive Moore, write in order to have something to read.

Real farming, old style.

Now that other farm, of about 184 acres, at Bellbridge, twenty miles away—the farm that saved my grandparents and aunt, and that is now farmed by a very sound man named Kincaid, whose father had it before him. (I may add here that my great-grandfather's uncle bought it from the heirs of a veteran of the first American Revolution, who had received it as a bonus from the state of Virginia, in the Virginia Military District.) To me, despite all this other history and sentiment, and despite the few hundred dollars a year that I get from it, this farm means chiefly three things: first and foremost, Mr. Kincaid himself; second, a beautiful place to visit; and third, a good chance to learn more about farming. I shall return to Mr. Kincaid and the visits later. Here I want to look briefly at the farm, as such, and at its operation. The only warning I have to make is that I so deeply admire the farmer and everything he stands for—as opposed to thinks—that I find it difficult to be objective about the farming.

Of the 184 acres, about 35, back some distance from Paint Creek, and on low hills, are in a lovely beech wood and more or less permanent pasture; about 14 are in barnyard and in scrub, beside the creek; and about 135, divided generally into five fields, are under cultivation in the bottoms.

These last extend down into a long point or peninsula made by the creek. They are undermined by water, and are flooded, to some extent, almost every year. The farm originally had 196 acres, but twelve were cut off by the flood in 1913, and are now unfarmed "islands" of scrub, to which we still claim title. The floods of January, 1937 came right across the peninsula, digging huge new holes in an ominously straight line, and depositing, not rich mud, but wide banks of sand and heavy gravel, sometimes a foot deep. The holes were from three to ten feet deep, and possibly ten acres of valuable land were permanently ruined. Incidentally, a promising wheat crop, with expensive timothy seed in it, was more than half ruined, and many more acres were covered so deeply with sand that it will take many years to restore their organic content. The County Agent said that my report to my sister, that the productivity and value of the farm had been reduced a third, was conservative. The water came over in three major waves of brown, foaming destruction, and reached the floor of a corn crib with several hundred dollars' worth of corn in it, and feeble foundations. With hundreds of people killed elsewhere, and hundreds of thousands driven out of their homes, neither Mr. Kincaid nor we could take it too hard, but the sight of that farm, after the water ran off, was and still is definitely nauseating.

Of course, we have the counterpart of floods, severe droughts, and, incidentally, the smaller, almost annual floods leave us a lot of seed of a dangerous pest called Johnson Grass.

On one boundary there is a levee at right angles to the creek, designed not to prevent flooding, but to check the force of the current when it is out of its banks. Its effectiveness is, for someone not an engineer, hard to judge, and experts are not in complete agreement on the subject. In our case the water cut around the levee with savage force.

Mr. Kincaid went right ahead with his work, as usual, and it was up to me to see what help we could get for repairs and prevention. In knowledge, capital, labor, and equipment, this was obviously beyond our means. City people who, like that erstwhile liberal, Newton D. Baker, have deplored the so-called passing of rugged individualism among our farmers and others, have never, apparently, encountered nature in some of her moods. I made a nuisance of myself to every government employee who might possibly bring us help, and finally, with the help of our true ally, the County Agent, I went up to the farm accompanied by him and his assistant and by the local chief of the Soil Conservation Service and his assistant, who do their work through the C.C.C. All these men were obviously sound, and on to their jobs, and wanted to help, but were faced by something much bigger than themselves, their C.C.C. boys, their equipment, and the money some people think is being thrown away. They had heard of the damage at our farm, and thought, perhaps reasonably, that they could use their labor to better effect on smaller streams, farther upland.

But they listened, and came, and tramped all over the farm and along the banks, and pondered, and debated, and finally decided that by widening the creek-bed at one point, and setting out trees in the holes, they could do something, if we should contract to run a band of permanent sod across the farm around the holes, and not to farm the upland more intensively to compensate ourselves for the income lost in this conservation measure. They would get in touch with me by a given date to confirm this and arrange details, in order to do the work in the summer. That date has not yet arrived, but when it does, I shall once more become a self-appointed Special Nuisance to Public Officials. Mr. Kincaid did not want me to contract under the A.A.A., and the Conservation Act, for reasons that seemed good to me, and, for that matter, to the County Agent. He now recognizes the need for help, and if I can figure the contract out in terms of dollars and cents, I think he will not object.*

Meanwhile we can learn nothing certain about the progress of the Conservancy Project, and the proposed Ohio Valley Authority. One hears rumors and counterrumors, and almost nothing seems quite so difficult as to find out exactly what is holding these measures up, and why. As I have said, state politicians and army engineers *seem* to be our chief difficulties. Many thousands of farms have suffered severely in this way, and our troubles are significant only in that they are typical. They say a dam has been planned and the site surveyed, a few miles upstream from our farm, but this detail, too, is shrouded in bureaucratic and political mystery. However, I much prefer the New Deal fog to the clear and glaring indifference of the Great Engineer and his immediate predecessors.

The business arrangement with Mr. Kincaid, which, so far as I know, has not been changed for thirty-five years, and which is a contract by word of mouth, is fairly simple. We pay the taxes, all expenses of upkeep of buildings and fences, insurance, half the cost of fertilizer, and half the cost of seed, except the most expensive seed, for hay, which we buy entirely; we receive half the grain in granary and crib, and all the hay, in barn. Mr. Kincaid supplies labor and tools, pays for half the fertilizer and seed, with the exception noted, and gets half the grain and all the corn fodder. We sell the hay to Mr. Kincaid, but he has to feed it, and all the fodder, on the place, which gets all the manure. The farm has a barn (with granary and stalls), a hay barn holding fifteen tons, with racks for feeding under cover, and a

*Later bulletin: The government boys are now very actively on the job. An engineer, an agronomist, and a forester are collaborating on a program that seems to me the chance of a lifetime. The only difficulty now is the psychological abyss between these young scientists and Mr. Kincaid, an abyss over which I walk a slack wire, smiling and beckoning towards each side.

corn crib. The old tenant house, which seems to have been a beauty, had to be torn down many years ago. Mr. Kincaid lives on his own adjoining place, which has about four hundred acres, somewhat less good than our one hundred and thirty-five in quality of soil. He has two brothers and a nephew, and all four work on their farm and on ours, under Mr. Kincaid's direction. They all work incredibly hard, and they have to hire additional help for husking corn, threshing wheat, and other jobs.

For many years, the rotation in the bottoms has been the familiar one of corn, wheat, and hay. Timothy and fertilizer are sown in a drill with the wheat, in the autumn, and red or sweet clover is broadcast the following spring, over the young wheat. This hay is harvested the following year, and the year after that, corn is sown again. This rotation is not considered destructive. Some farmers in this region don't have to use commercial fertilizer drilled with wheat. Some have to, but don't. Some alternate between corn and wheat, which are both soil-destroying, without even putting back manure. When I investigated payments under the Soil Conservation Act, I found that I could not collect on our more or less permanent upland pasture, because we had not farmed it within five years, and as I have said, at that time, before the flood, the County Agent did not advise us to withdraw, and put in leguminous hays, any more of the bottoms. Mr. Kincaid feeds all his hay and fodder, much of his corn, and sometimes even wheat, to about forty head of cattle (which he does not milk), many head of hogs, and some horses and chickens. We have to sell our grain, and the corn (except for a little hauled to Oak Hill) generally goes to a neighbor who specializes in feeding hundreds of head of livestock.

The reader will have noticed that these two farms, the Kincaids' and ours, taken together, illustrate almost perfectly the principles I have tentatively put forward. The uplands are slightly, but not dangerously, eroded, and nothing but fully grown timber not in strategic positions is ever sold. I don't know what Mr. Kincaid's income is, and I don't consider it any of my business, but I do know that he spends money, his or mine, only for upkeep and as a very last resort. His thrift and ingenuity are really marvelous. Furthermore, he doesn't sell anything but unquestionable surpluses of livestock, that can't possibly be used, or put back into our farm or his. He accepts the markets with the cheerful, laconic fatalism with which he accepts droughts and floods, and is not at all interested in the Coöperative movement, or in any other social possibilities or implications, but for all I know, his nephew may be. If you don't look beyond your own line-fences before you're forty, you never will.

Under the present arrangement, I find that I have little or nothing to criticize or suggest. The only real contribution that I have been able to make to our partnership, so far, is in passing on to Mr. Kincaid some very useful "grain letters," on the state of the markets, that my brother-in-law gets from Chicago and sends on to me occasionally from New York. I had some soil samples tested, and sent the results to Mr. Kincaid, suggesting that he try some alfalfa. He was hesitant, but it looks as though the flood, and the possible contract with the Soil Conservation Service, would get me my alfalfa at last. It has become almost a standing joke between Mr. Kincaid and me. I have constantly urged more fertilizer with the wheat, especially in certain places, and Mr. Kincaid has seen this work. I go up every week or fortnight, and we do quite a lot of talking, but of course I get much more out of these conversations than Mr. Kincaid does. I can imagine more modern uses of the farm, but I am not sure that I can imagine better ones. And I am quite sure that I can't imagine a better tenant. More even than his competence, economy, hard work, vigilance, and all the rest, I value Mr. Kincaid's utter honesty. He is in a position to cheat me right and left, and I'd trust him with my last nickel, my last shirt.

Neither, under our present economic system can imagine a better working arrangement. I'd like to feed our corn and hypothetical alfalfa to Herefords, and to sheep, respectively, on the place, but Mr. Kincaid has neither time nor strength to feed livestock of mine, and I am too far away to feed them myself. Besides, the summer pastures are chiefly his, not mine. If I were in Mr. Kincaid's shoes, I'd sell nothing but cattle, sheep, and wool, but then I'm not, and if I were, I'd make a lot of mistakes that he does not make.

However, I can think of a better economic system, which would give both Mr. Kincaid's farm and ours to the American people, which would test the full possibilities of the land, and use it more closely in relation to the public needs, both now and in the future, which would give Mr. Kincaid complete charge, under the general supervision of the Department of Agriculture, and pay him at least the equivalent of the present total income of the farms, which would substitute human foresight for a brutally capricious market, which would reforest and build that dam upstream plenty quick, and provide Mr. Kincaid with cheap electricity, and finally, which would quite eliminate private landlords, however admiring and friendly they might be.

Often when I go up to that farm, and get checks and good advice, all for nothing, and try to talk with Mr. Kincaid's brothers, who are so painfully respectful to the owner, I wish to God I were their and Mr. Kincaid's apprentice, working with them and getting paid for that, and getting respect, if any, not for my great-great-uncle, my great-grandfather, my grandfather, my grandmother, my father, and my aunt, but simply for myself.

Sine qua non.

In short, we constantly violate the principles we have learned, but we do so consciously, they still influence our actions strongly, and we still think they are sound. If and when we go bankrupt, we may have the not wholly grim satisfaction of knowing why.

There remain one final question, and one final principle, that are probably more important to the city-bred farmer, and to the small farmer generally, than all the others. The question will have arisen in the reader's mind already, perhaps in some such form as this: "This is all very well, and perhaps sound doctrine for thousand-acre farms, but has it worked in your case? Are you living on the income, if any, so derived? What about those 'few thousand dollars' you have mentioned? What about city people who want to farm, and who will not be thrown into a flying start as you were?"

I will reveal the concrete answers to these questions later, when I go into the details of the red and the black. Here I can say this: the farm at Bellbridge brings my sister and me each a few hundred dollars a year. Oak Hill is now run at a loss, which we have diminished steadily, and could eliminate by not having any hired help, and may eliminate with help. We regard this loss as a temporary rent, and remember that we could not get two good rooms, with bath, in Greenwich Village, for the same amount. The gruesome fact is that we need thirty or thirty-five hundred dollars a year to live in the luxurious style to which we are accustomed. Our first year, we spent a good slice of our capital; the second year, acting solely on expert advice in a rising market, we "made" enough money almost to stop capital expenditure; and this third year we are hoping to keep out of the red with my specialty, words, and are thinking of putting most of the rest of our money capital into further land and equipment.

The final answer, the final principle, I think, for small and new farmers, is a specialty, a trained and marketable ability that can be combined with farming. It is this problem of the "cash crop" for small farmers that keeps them, and Secretary Wallace, awake at night. I am thinking of raising only purebred sheep, in order to sell the bucks for more than lamb. Peggy has experimented with dogs, and is now investigating more likely chickens. We may some day run a sort of summer school. Some day before I get too old I may study veterinary medicine.

With one good hired man, on a place of this size, and with Mr. Kincaid at Bellbridge, writing is perfect as a specialty on a farm, except that the crop and cash both are notoriously uncertain. Like most writers, I have a wonderful time, with intermittent tortures of the damned, and would advise no one to specialize in anything so neurotic and hazardous. If I could exchange whatever talent I have as a writer for that of a trained and experienced veterinarian, I'd think a long time. In any case, an arty attitude towards writing, on a farm, is not possible. I think the arts, artists, and the public generally would all be sounder if all artists were able and forced to sell their products to their neighbors before thinking of New York and Hollywood. I tried to get on the local branch of the Writers' Project, in the W.P.A., and found that I could not do so until I had been on relief. That seemed fair enough. I tried to sell a weekly column of comment on local affairs to one of our two newspapers, and was willing to take what I thought was the lowest reasonable sum, five dollars a column. The editor and publisher turned me down, and told someone else he saw no point in paying me that much when he could get Walter Lippmann, canned, for two. That seemed fair enough, too, though it smacked of technological unemployment for me.

In any case, a specialty is necessary, or else a few thousand dollars for a period of adaptation and reconnaissance, during which one might possibly be acquired.

IV. Production

Varieties of religious experience.

I understand that every spring, in Plymouth, Massachusetts, a little procession of people, in Pilgrim costume, commemorate an event of three centuries ago by going out and planting some corn. For fertilizer, a fish is planted in each hill, with the seed. The incumbent Minister of the First Church says a prayer. Now I don't happen to go to church, or to pray. To me, all that is a confusion of fear and thought, expressed in a poetry that is too cold, respectable, and timid, and often too vulgar, in the wrong way, to be capable or worthy of expressing any real thought or real emotion about life and death. Still, every year ten million other farmers, and I, in this country, do certain profoundly strange and important things, and sometimes have strange thoughts and feelings about them, and this little ceremony in Plymouth is the only one I have heard of that recognizes, in any way, their religious significance. It's probably just as well, because all the clergymen, with the exception of this one, would doubtless try to take over this Act, too, and make it as cold, respectable, timid, and vulgar as their christenings, marriages, communions, and burials. I'm sorry I mentioned the matter at all.

The first preparatory rite is hauling out the manure, in autumn or winter or early spring. My hired man, James, and I are able to handle ours with my little truck. It is tiring, but not half as nasty as it seems. Good manure is a lot cleaner, for instance, than street refuse. Ours rots all summer.

Then the plowing, dragging, and harrowing. I have tried several times to plow, years ago as a boy, and recently, and I have never been able to do it. If I had more chance, I think I could learn, but so far, every time, the plow has gone too deep, or jumped out of the ground, or wandered all over the place, or all three, and after about five minutes I feel as though I'd run a mile. Dragging and harrowing are easier. I don't know anything much nicer to look at than a good field all ready for seed.

We have planted corn here twice, now, in my tenure, but I do not know how to operate a corn planter. It seems to me a very strange and inefficient machine. A long wire, with pieces of metal attached to it at regular intervals, the distance between the hills of corn, is stretched across the field. The planter is driven up the wire, which releases the seeds of corn. The wire has to be moved every time a new row is begun. I should think that a catch on a wheel would do the trick, but then I'm no mechanic. It is hard to believe that these odd little machines being driven along wires in May are planting seeds that in a few weeks will be square miles of corn as faultlessly ordered as a drill at West Point, and much more exciting.

A wheat drill is larger, more complex, and more ingenious. It is used in the autumn, here, to plant wheat and other grains, and at the same time it plants timothy seed and drills in commercial fertilizer. We have had to plant rye twice, here on the hill, as a cover crop for hay, and for early pasture itself. The first time, I called up a neighbor and arranged to borrow a drill. My hired man at the time, Hubert, went and got it, with his father's team. It had been loaned to another neighbor, who had allowed fertilizer to get wet in it. Now fertilizer hardens like cement.

Hubert and I spent a morning cleaning it out, and then Hubert got to work with it. I followed for a while, and everything, to my ignorant eye, seemed to be working properly. When he got through, Hubert reported no further trouble. It was only the following spring that I found that all the timothy seed (eight dollars a bushel when bought) had been dropped in a small area at one end of the field, and a band twenty feet wide, the length of the field, had not been planted in rye. I have rarely felt quite such a fool.

The following autumn, after the other field had been cleared of corn, and Hubert had departed, I got a neighbor to harrow it with a tractor, and I sowed the rye and timothy my-self, by hand. I used an old seeder which is simply a gadget at-tached to the bottom of a sack. You fill the sack with your seed mixture, open a vent at the bottom, turn a crank that turns a disk that throws the seed many feet, and walk steadily down one corn shock row after another. After a couple of rows, the seeder broke down, and I had to finish the job by hand. I felt certain that I could not spread that seed evenly over thirteen acres, and make it come out right, but it did. A couple of weeks later, when I happened to be in hospital, Peggy reported that the rye had come up evenly all over the field. That winter, most of it was frozen out, and it gave us only a few days of pasture, but I'd rather be beaten by twenty-two degrees below zero than by my own ignorance.

Then came the question of putting in spring grass seed. A little timothy had survived, but not much. Ordinarily, one sows clover at this time, as I had done in the other field. But I had noticed that the clover had only done well on the slope that was drained. I consulted our County Agent, and he advised a mixture of timothy, red clover, red top, lespedeza, and Ken-tucky blue grass. I put in about twelve pounds of the mixture to the acre, and bought enough to reseed part of the other field. I also bought a new seeder.

A couple of washtubs full of grass seed are as exciting as, say, a publisher's stock room. I plunge my arms into that cool, clean seed, and look at those mysterious, hard little grains, and I see thousands of rolling acres of grass, with roots gripping the earth, and the wind caressing the green, and great herds of cattle and sheep eating it , and lying down in it to rest and chew, and growing fat and strong. I hear the grasses murmuring, I feel them drawing up the water and the richness . . .

Then, on a clear warm day in March, without wind, when the pools of water had mostly disappeared, but the earth was still damp, I did my sowing. It's an easy job. The sack isn't heavy, and I sowed the thirteen acres and more in less than a day. The seed flies out like hard rain, and stings your hands red. You walk steadily over the earth, in the sunlight, looking up, now and then, at the hills. You imagine the grass growing. It doesn't have to be cultivated, and with any kind of soil and dampness at all, it spreads. It's good food; it's a fine, useless beauty, too. You think of Johnny Appleseed, moving west in this very country, when it was a lovely wilderness, and Abraham Lincoln's father was chopping down trees. You think of the Mongols and their herds, always moving, always looking for grass. You wonder whether, if you do something like this, the figures in your book are so terribly important after all. You remember a farmer's telling you how his father sowed grass on horseback. You think of getting your animals through the winter alive, and you know that from now on the earth is going to help you, and them, free, gratis, for nothing. You want to sing, dance, yell, get drunk, and pray. And you walk on, steadily, listening to, and watching carefully, the fall of that divine Rain.

Like most religious ecstasies, this one is followed by doubt. Is the damned stuff going to grow after all? Was that clay too wet? What about those pools of water that collected in the next few days? Then the tiny seeds begin to sprout, and take root, and you see that they are doing much better in some places than in others. And the weeds appear, too. Then it doesn't rain for three weeks, and when you wake up at night, you listen for drops on the roof. But then one day, after the stock have eaten off most of the rye, which has a different color and texture, you are walking out across the fields, with dogs, and you notice suddenly something new, a fine green fur, not the old rye, not the weeds, not moss, but grass! . . . Mine, last summer, was almost completely wiped out by a four months' drought. This year, I got a better start, but so did the weeds.

But the next spring, after another harrowing, I went out again with a sack of green mystery. I think you do this year after year, on clear, warm, windless days in March, whether it makes any money or not, whether you can afford it or not, and whether you get any breaks in weather or not. "There goes that old boy," they say, "still sowing grass." And then one fine day it's all over for you, and they take your body, like the fish at Plymouth, and put it in the earth. The process is fancy and silly, but in a short while all that, and all you, are quite forgotten. But meanwhile the roots of the trees and the grass are reaching down into the earth to the ashes, to the skull, which feeds them, and the leaves and grasses are waving in the wind and sunlight.

Invitation to a science.

I don't possess an orderly and scientific mind, but I am much interested in what might be called the scientific background of agriculture, and I hope that some day there will be closer and stronger connections, through alert and curious farmers and more practical scientists, between farms and laboratories. It may be that as more people do more or less what I am doing, some of them will be better equipped for this work of liaison. Whether they have the necessary minds or not, I can predict that they will have plenty of fun.

Of the sciences related to agriculture—and few are not related to it, in one way or another—the one that interests me most is genetics. Not since I studied elementary structural geology, fifteen years ago, have I encountered any science that seemed to me to be so closely related to—what shall I say?—to the critical points in the situation of man on this planet. But a few warnings may be only fair play.

It is an extremely difficult science, or at least, I find it so. I have studied the elementary books by H. S. Jennings and Sewall Wright off and on for several years, but right now I should have to turn to the books before I could undertake to explain such relatively simple points as sex-linked and Mendelian characters, with accuracy.

Also, and this is even more important, laboratory animals, like the celebrated Drosophila melanogaster, and guinea pigs, can be bred and studied so much more quickly, easily, and cheaply than barnyard animals and garden vegetables and flowers, that the great advances of the science in the last few years are, to a farmer, more exciting and challenging than helpful. It is one thing to understand, after hard study, the basic phenomena and theories; it is quite another to relate these to domestic animals and plants, and to breed what you want. For instance, no new breed of sheep has been developed in the United States. Dr. J. McI. Phillips, of Galloway, Ohio, who is probably the most scientific breeder of cocker spaniels in the country, tells me that it has taken him thirty-five years to discover with certainty what matings will produce red cockers.

During a couple of winters or so in New York, some years ago, I became what is known, with appropriate disrespect, as a "Saturday afternoon expert" on a certain limited kind of prints. The joke was that by the time I got through, I actually did know more in that field than any but perhaps a dozen real experts. In breeding and genetics, there are the scientists, and then the farmers, who are too busy to do anything more than guess, and look at the purebreds at fairs. In between, there may be, as I hope, more scientific breeders, with endless patience, curiosity, skill, and scientific intelligence. But there will never be any Saturday afternoon experts.

But we immigrant farmers can keep our eyes and ears open. One thing I want to see, before I die, is that movie I've heard of, of the union of sperm and egg. All history, all fate, is in that moment, and when I see it, I fully expect that some wrathful god will strike me blind on the spot.

Meanwhile there are certain repeated little episodes that suggest, not the professors and research students, but rather Mr. Pickwick, or Mr. Shandy.

"They order this matter better" in laboratories.

Early one morning a few weeks after I came here to live, and when I still had ink on my fingers, Hubert Brooks, my hired man at the time, seized me at my solitary breakfast and announced in excited whispers—for Hubert was nothing if not delicate about these matters within earshot of his wife and children—that Daisy had "come in." Daisy was a perfectly ridiculous long-legged gangly red roan cow, vaguely Shorthorn in blood. Hubert was plowing a field, and couldn't very well leave his father's ancient team standing still long enough to take Daisy to a bull. Extremely slow motion, in a hired team, is better than none. I am not very keen before breakfast, but it was clear enough that it would be well if I escorted the impassioned cow to an assignation. My aunt had used, and we were using, a good beef Shorthorn bull, all red, owned by our neighbor to the south, Ralph Stone.

I was rather irritated, because a few weeks before, I had written to another neighbor, a rich old gentleman, who had been a friend of my father's and aunt's, asking him, quite brazenly, whether I couldn't breed my cows to his purebred Hereford bull. I wanted better bull calves, and I remembered that this neighbor had for years kindly allowed my aunt to send her cows to an earlier champion of his. There had been no reply. I had already calculated that if I wanted to keep the supply of milk steady, I should have to breed Daisy on her next time in, regardless.

I sent Hubert back to the plow, finished my breakfast, got a rope with a nose-ring, caught the wide-eyed matron, and started south through Stone's orchard. A cow stays in heat about a day, but there's never any telling how long she's been in before you noticed it, and there isn't any time to waste. The cow, at least, always actively concurs in this opinion.

Well, I walked miles and miles and miles with that cow on the end of a rope. Someone would tell me the bull was in such and such a pasture. Laboriously, Daisy and I would make our way thither, get through a gate, and ransack every thicket, before finding a herd of . . . cows, calves, heifers, and steers, all very curious and sympathetic, but definitely useless. Finally, along about noon, we entered a perfectly enormous field, leagues from home, and saw a herd that included the bull. At least, I thought he was the right bull. He was a huge beast, all red, and well built. I wasn't absolutely sure that he was the right one, but at that moment he looked good to me, and to Daisy, and we were both prepared to face any consequences of a mistake.

Boldly we approached the harem, and rather quickly the gentleman got the idea. It grew on him, and he began to paw the ground, snort, and reconnoiter. I thought it would be a good idea if I merely dropped the rope, and left the ring in Daisy's nose, so that I could catch her afterwards. The nearest fence was a hundred yards away, and I had never been recognizably a sprinter.

Soon enough, my error became apparent. In their ardor, Daisy and her friend had become all tangled up in that rope, unbelievably tangled. As far as love went, they might as well have been Pyramus and Thisbe. They began to wrestle painfully with adverse fate. The gentleman, especially, was very much annoyed. I had not made it a habit, in my short and merry life, of unwrapping packages of cows and bulls, but there wasn't anything else to be done, and I did it. In the process, I took the ring out of Daisy's nose. Love conquers all. I wiped my brow and lit a cigarette.

I didn't know how many times . . . how many services were advisable, but after all that, I didn't propose to be niggardly. Finally I decided that enough was enough, and began to pursue Daisy with my rope and ring. She didn't like the idea, and neither did her friend, but I finally succeeded, and gingerly began to lead her towards the distant gate. Parting is such sweet sorrow that the gentleman escorted us all the way across the field. He wanted very much to come right along home with Daisy, if not with me, but I finally nipped that idea in the bud. An hour or so later, exhausted, and plastered with mud, we reached home. I released Daisy, who had calmed down a little, and made a sprint to the ringing telephone. It was the farm-manager of the owner of the Hereford. He said I could send my cows up whenever I wanted to. I thanked him and hung up.

In the interests of science, I may report that the offspring of this union was a heifer, all red like her father, but long-horned and gangly, like her mother. No, not gangly; that doesn't do her justice. She became vaguely suggestive of a gazelle, really quite pretty, but from my point of view, absolutely no good. I wanted nice, chunky little bull calves. I sold her as a yearling.

But I got what I wanted. From this Hereford bull—a low-slung, enormous beast, as solid and square and pure of line as if he'd been hewed from stone by, say, Mestrovic—we have had five calves, three male. In build they aren't purebred show animals, but they are good and solid, and the best we have had on the place for years. In marking they are definitely Hereford, but one of their white faces is so ludicrously splattered with red that it looks like a clown who has made himself up while as tight as a tick. I wanted somehow to express my gratitude to the owner of the bull. When my aunt was confronted with the same problem, she sent him one of her huge hydrangea plants for his garden. All I had to send was a copy of one of my novels, with regrets that my mind as not as fertile and prepotent as the Hereford. I do not expect this favor to be continued indefinitely, but the neighborhood now has a fair number of good Hereford bulls, and the general quality of cattle has improved greatly within my own relatively short memory. Our farmers have discovered that good sires pay, in dollars and cents. If the sheep do well, we shall try—as I have said—to save winter feed and have better milk by keeping only two good Jerseys, and selling the calves for veal. Even so, I shall continue to breed to Herefords, because except for heifers for replacements, Jersey calves are no good, while calves of this crossbreed bring almost top veal prices.

This present Hereford bull—long may he reign, and may his tribe increase—lives a couple of miles down the pike on a large and well-kept farm, with a younger male colleague or two, and a herd of some forty good, grade cows, originally founded on a shipment from Kansas City. Our parades down the pike to this gentleman are less ludicrous than my adventure with Daisy, but they are not without their lighter, more desperate moments. In these matters Hubert always showed a refinement and reticence in which I am notably lacking. He would never breed a cow right by the pike, with the traffic going by, but would always lead her back from the road, to a clump of trees, or over a knoll.

Once when I took a cow down alone, the affair became almost a public scandal. I found the bull, with the other bulls and all the cows, beside a gate right on the pike. This gate was chained and padlocked. The house where I could get the key was a mile away, I didn't want to walk the cow all that way and back, and I couldn't leave her tied where she was while I walked two miles. I tied her to the fence, and rushed to the telephone of a friend living nearer. I wanted to get Peggy to drive like mad for the key. The line, of course, was in use. When I finally got it, there was no answer at home. With the big bull on the point of wrecking the gate or fence or both, I couldn't wait. When I got back to my cow, I saw, not far away, a young town boy selling watermelons and cantaloupes at a roadside stand. He had a car. He was curious, and then embarrassed, but I got him to drive for the key, and promised to mind the stand during his absence. While he was gone, I moved my cow to the other side of the road, some distance away, but all the bulls were working themselves into a lather. A man came along and bought a watermelon, but I didn't have any change, and had to give it to him, and pay for it myself later. When my friend in need got back, he said the farmer said the key was in such and such a crack in such and such a post. We both looked, but couldn't find any crack or any key. At this moment, another neighbor came along and wanted to know what the matter was. I told him, and he told me of another gate, that I had never seen, some distance around the corner of the field. Some time later, three different town friends, who had passed in cars that morning, wanted to know whether I had raised all those melons I had been selling. In all, this little affair had to be explained to thirteen people.

Rams, and albatrosses.

As I have hinted, one of my current daydreams is of raising purebred Shropshire sheep. So far, sheep have been the most economical and appropriate animals for this place, and except for the dogs, I like them the best. Purebreds take no more care than crossbreeds, which they can replace gradually, so that a lot of money need not be gambled at once. Shropshires are the most popular sheep in this part of the state, but to my knowledge, there isn't a purebred flock of them within forty miles. If I found no market for them as flock sires and dams, I could always dispose of them through the regular sales, and have that much more wool, that many more lambs, merely as wool and lambs, regardless of blood. It certainly seems to be true that a good ram is a better investment than a good bull or a good stallion, because he costs and eats less, is easier to take care of, produces more offspring, and takes more effect on them. Sheep men seem to keep their eyes on market, rather than show values, in breeding. There seem to be plenty of bulls and cows that are what Secretary Wallace calls "registered plugs," but I have not yet heard that criticism of purebred sheep.

When I made my first purchase of seventeen ewes, fairly good ones, but definitely crossbreeds of the mutton types, at $8.00 a head, I bought also, at the same price, a grade ram, chiefly Hampshire. He had a slight sag in the back, but he looked good. He turned out better than I had thought, because the percentage of lambs was 135, and they were all better than their mothers. The next year I bought a purebred Shropshire ram from a breeder forty miles away, and paid $25 for him. He was better than any Shropshire ram I saw for sale at the state fair when I got there late in the week, to be sure—and $25 was the bottom price at the fair. When my ewes got through with him this last fall, I loaned him to a neighbor. This spring it appeared that he had been an excellent investment.

I find it hard to learn to judge sheep, because those at the sales aren't generally worth looking at too closely. They had a wonderful lot of good sheep, of at least a score of breeds, at the state fair, and I spent hours looking at them, studying the wool on exhibition, and watching the shearing contests. But I can see that it will take a lifetime to learn enough. Meanwhile, I look at every sheep I see, read the monthly *Sheep Breeder*, and go back again and again to W. C. Coffey's *Productive Sheep Husbandry*. The leading breeders are still importing from England, and the signs above the sheep at the fair were a roster of the English shires. Looking at those sheep, and reading "Shropshire," "Leicestershire," "Southdown," "Dorset," "Cotswold," "Lincoln," and so on, there in Columbus, Ohio, I could almost hear the skylarks. Then there were the Rambouillets, and the Merinos, with their own auras of memory. The irony of it is that when I was in Europe, I couldn't tell one kind of sheep from another. I did spend an hour or so watching a sheep-dipping on a farm near Wordsworth's and Coleridge's Nether Stowey, and talking, to my delight, with their owner, but I couldn't say, right now, what kind of sheep he had. At that time I was thinking more about albatrosses.

Postscript on pedigrees.

My aunt used to have the greatest mixture of chickens known to man. Their incongruity amused her, and she said the crossbreeds were healthier; I think they were, in the first cross, but not later. For two years we had more or less pure Rhode Island Reds. These are large, good to eat, and lay a fair number of good big eggs. However, we found them very unhealthy, and after talking with others, attribute this in part to a weakness inherent in the breed. We shall probably go into one of the crosses, and keep buying chicks.

Our two geese, Philander and Arabella, are of the familiar gray and white Toulouse breed. Arabella lays a large number of eggs, in fall as well as spring, but has hatched out few, and raised none, though at the present moment she is setting, and once more we are hopeful. She is not naturally a homebody, and yields too easily to the raucous persuasions of Philander to leave her nest for strolls and swims. Neither has she had the best coöperation from us, and on occasion has been forced to lay her eggs in the dog kennel.

Our ducks are White Pekins. They are noisy, healthy, and excellent for the table. They have reproduced themselves adequately, if not brilliantly. One year I forgot to get a new drake, which was not disastrous, though the inbred ducklings were not so strong. If duck eggs were more edible and numerous, I think chickens would practically disappear. Duck eggs are stronger, but Peggy has used them successfully in cooking.

Two of our cocker spaniels, Jeanne and Peter, were given us by a former boss of mine who has bred and judged cockers for years. He also gave us Peggy's first dog, which was killed. Our third cocker, Geoffrey, which we bought, looks and acts like an excellent mutt. All three have pedigrees and numbers a mile long, and their ancestors are pictured in all books on the breed. When we bred Jeanne and Peter last year, we got two good dogs out of five, and shall not repeat that mating. My aunt's dog, Jack, is largely collie, with some mixture of a breed that is fairly definite in this region, though unrecognized. We call them shepherds; they are like collies, but stronger and smarter, long-haired, black with brown markings, and a good deal like what sheep men call border collies. Not only for sentimental reasons, I prefer Jack to any registered collie I have known. I hope that some day the whole matter will be clarified, and collies restored to their original strength and intelligence.

All four of our dogs are individuals, and our closest companions, so they will naturally reappear in these pages in other connections than the genetic. Peggy knows much more about dog breeding than I do, although that isn't saying a great deal. Still, my opinion on one point has been asked repeatedly, it is fairly unprejudiced, and it can be briefly stated: There are plenty of superb mongrels, and "papers" are often sheer snobbery, but it seems to me perfectly true that if you know what a breed is like, in general, what the main blood-lines produce, and what the ancestors of a dog actually were, as dogs, rather than as names and numbers, you are on safer ground, not only for general appearance and show points, but also for health and character.

One hundred clamoring stomachs.

Within the general principle of raising the kinds of animals that your place can feed, and that can feed each other—that is, again, of buying as little as possible—each kind of animal, at each stage of development, presents a separate and interesting problem of feeding. Each farmer has his own favorite notions on these matters, and state and federal agencies provide free hundreds of reports of experiments in feeding. These experiments frequently, though not invariably, ignore the economic element in the problem, and so the pamphlets have to be used with extreme care. Once you have the right animals, in the right numbers, for the feed and land you have, and for each other, it is better to risk an unfinished or second-grade (if healthy) product than to risk deficits caused by feed bills. For what I may call immigrant farmers, like ourselves, the temptation to produce the best animals, and get the top prices, at any cost, is likely to be strong and dangerous.

The actual feeding of all our animals is hard physical work, but it is much more fun than all this calculating and speculating.

In the winter, our man James has a good deal of housework to do in the morning and evening, so unless I am too busy, or away, I generally do the milking and feeding. Eight or nine head of cattle will clean up a shock of corn fodder in an ordinary winter day, and another half or whole shock on a very cold day.

A good shock of fodder weighs, I guess, several hundred pounds. Sometimes it can be lifted or carried in smaller bundles on a fork, but the only way I can do it is with my arms. Ordinarily it is hauled in from the field, after husking and reshocking, and set up for the third time near the barnyard. James and I hauled some in my little truck, but then the fields got wet, and we had to carry it to the barnyard, through mud, or to and over a fence where the cows could get it. This is hard work with a vengeance, but not unpleasant. I used to do it mostly in the late afternoon. The dogs always come along, to hunt field mice in the shock-bottoms, and winter dusk is often beautiful. All I had to do to enjoy it fully was to remember winter evenings on commutation trains.

Corn fodder is filling, and gives cattle something to chew on all day, and they can be "roughed through" on it alone, if you have enough of it, but through necessity and choice I have generally fed some hay and corn. Naturally, cows carrying or nursing calves, or merely being milked, and steers being at all fattened, have to be fed up rather carefully. Hay is much easier than fodder to handle, even when it is fed out of a rick and over a fence or two, in wind and snow. In the summer I wean calves on grass, with maybe a little bran and cracked corn. In the winter, a calf being weaned takes a little pampering, or a fight. One winter I had a calf who refused absolutely to eat anything but his mother's milk at the source. We tempted him with warm milk, mashes, and everything we could think of. He got very thin, and Peggy begged me to give in, but I didn't, and he finally began to eat with interest. However, this little hunger strike set him back perceptibly. Once I fed my cattle some cottonseed meal, but I don't think it paid. At the end of this last winter, with hay very expensive and the grass very slow to come on, I fed straw with molasses, and corn at milking. This was not a real economy.

Sheep don't eat nearly so much as cattle in the winter. Last winter I had twenty-five head growing too fat on not a great deal more than was being eaten by two cows not growing fat. Unlike cattle, they will forage through snow on the ground, if there is anything under the snow—as there is not after a severe drought. I feed them fodder or alfalfa or clover or a little corn. My best hay is saved for them in the last few weeks of carrying lambs and during the first few weeks after lambing. At this time I supplement the hay with a grain mixture. I probably spoil mine, but I think it's important to keep them gaining through lambing. They say too much corn makes a sheep lose her wool, but I think bad breeding and bad physical condition generally are more likely causes.

After lambing, I used what is called a creep, that is, a pen with an opening large enough for lambs, but not for their mothers. In this, for some weeks, I kept a mixture of cracked corn, bran, and oats, and it pays.

Both cattle and sheep require a constant supply of salt, and for the cattle I keep it standing in their feed racks in fifty-pound blocks. I feed my sheep a somewhat more expensive salt, that is medicated for worms, and also rich in minerals that upland grasses are apt to lack. Whether it is worth the extra cost, I do not yet know.

We feed our pigs corn, skim milk, extras from the garden, our own garbage, and garbage collected by James every other morning from neighbors on the road.

All this isn't the best feeding, but it's the best I can do. In feeding chickens, I haven't even begun to work it out successfully. I feed chicks a dry "starting mash" that has to be bought, but that is supposed to contain every essential, and does actually work. Then they get finely cracked corn, and milk, and finally, shelled corn, or corn on the cob, skim milk, and garbage.

I have used laying mashes with results, but these are expensive. Also, at least half the time, they get the run of the place, with, in the summer, green grass, insects, grain waste, and everything else. I also feed oyster shells, which harden the egg shells. Now this diet is supposed to be good, but has not proved so. My chickens have been thin, layed poorly, moulted early and long, and been attacked by all known chicken diseases. We try to keep buildings and utensils clean, but finally had to sell out, whitewash everything, disinfect everything, plow up and lime the hen yard, and hope for some hard freezes. Now in the spring we shall start again.*

Our most flagrant violation of the "buy nothing" idea is in feeding the dogs. We have few scraps, and these are consumed by pigs or fowl. Besides, scraps don't make good dog food. Except for puppies, which require feeding too elaborate to report here, our dogs are on the following diet: breakfast, a few crackers with a little milk; dinner, at five o'clock, a pound of canned dog food for Jack, and a half pound for each cocker; bedtime, one cracker each. At least once a week, and usually oftener, they all get raw hamburg instead of canned food, and bones. We buy canned food by the case and crackers by the hundred pounds. Now and then there are spots of vegetables or corn bread. They all hunt avidly, but don't generally eat the rabbits, moles, rats, and mice they catch. However, I once was unable to keep Jack from swallowing a baby rabbit alive and whole, at one big gulp, and it did him no harm. Incidentally, psychologists investigating the experience of time would do well to look into animals. Much of the time, more than half our clocks aren't running, but we always know very well when it is five o'clock.

*Later bulletin: The infection had not been removed.

Milk pail rhythms, with accompaniments.

When you have only two or three cows, milking them is not a hard job, but it is the core, somehow, of all the morning and evening chores. Milking, of course, has to be learned, but it is like swimming, or riding a bicycle, in that once the knack is acquired, it is never lost. I learned how to milk one summer when I was about fifteen years old. My aunt had no help, or at least outdoors help, at the time, and I milked the four cows. My hands, wrists, and forearms were weak when I began, but much less so at the end. In college, I did some mediocre rowing on 150-pound crews, and learned how to play squash racquets, which I kept on playing more or less regularly until I moved to the country. The result is that I do not now get tired when I milk, and my wrists are the only moderately strong parts of my body. In his autobiography, Bliss Perry says that the milking he did as a boy strengthened his wrists for fishing, but alas, I am no fisherman.

I had a lot of fun, in the barn, that summer years ago. I was very proud of my job. I had a cat and three kittens that I taught to follow me out to the barn, line up in a row beside me as I milked, stand up on their back feet, and catch the fresh milk that I squirted into their mouths. They had to drink like mad, and even so, a good half of the stream was spattered on their faces and bodies. After each had had a shot or two, they would all retire to corners of the barn and spend a delicious half-hour licking themselves off. They loved it, and so did I.

Jack was never taught to fetch the cows, and so he still gets too excited, and runs to their heads. However, he is far from useless, because when I go to fetch them, and they are scattered, or become so, he can always get them started, and once they are started, they usually head for the barnyard. One reason he doesn't drive cows properly is that I have never taken him on a long rope and taught him, and another is that my aunt taught him to run at the cattle, barking savagely, whenever he heard them leaning on fences. He loves it, and at the creak of a wire, he is off, with all the cockers trailing and yapping behind him. It doesn't even take the creak of a wire. On summer evenings we usually sit in deck chairs on the front lawn, with the dogs beside us. The air has cooled, and they are all eager to go. If we have a calf alone in the old orchard adjoining the front yard, it takes nothing at all to send the whole pack after it. Our latest calf being weaned developed such admirable nonchalance that he rarely moved more than a few steps, and of course this passive resistance quickly dampened the ardor of the dogs. In fact, all our cattle are so used to the dogs that they can rarely be broken into a run. A friend of mine who used to be a cowboy tells me that the ranchmen out West wouldn't tolerate dogs, except greyhounds used only in coursing coyotes. However, I doubt whether all our canine alarums and excursions do our cattle much harm.

Of the cockers, Peter and Geoffrey like to go with me or with James to fetch the cows, but only Geoffrey has a passion for the job. He is learning how to drive slowly, from the rear, but he is still rather ridiculous, because he is such a small morsel of dog behind the cattle, and when they turn on him, although he rarely yields ground, he usually tries to pretend that he is admiring the view. Cocker spaniels aren't supposed to be working farm dogs, but if anyone took more time and trouble with them than I do, I don't see why they couldn't work very well.

I always like to fetch the cows, because it makes such a pleasant walk at the two best times of the day. The dogs always enjoy it so much, and so did Peggy's small niece, when she was visiting us. I like to see the condition of the pasture, and to make mental notes of odd little jobs of fence repairing, or weed cutting, or brush gathering, waiting to be done. There is always something nice to watch, like a rabbit making fun of the dogs, or a covey of quail taking off like a squadron of planes, or an exquisite little skunk driving the dogs crazy, or a long black snake going his own mysterious way. One can see most parts of the farm from our yard, and if the cattle are nowhere to be seen, they are way down on the east hillside, below the old grave-yard, or hidden in the trees on the edge of the cliff to the west, or taking their ease in their shed. It is easy to forget this last possibility, take a long walk for nothing, and come back feeling silly. James is so much better at most of the jobs than I am that it always tickles me when he does something like this. One day last summer he went twice all the way down through the woods on the cliff to the west, to the railroad, and back—a good two miles of bad walking—before he found them in the shed.

Sometimes, of course, the cows come home at the right time all by themselves. During the drought last summer, they did this too often for comfort, because the pasture was so dry that they had to look forward to the corn we fed them at milk-ing. Then, too, they paid attention, for the first time in years, to a repeated call of "*Soo*-ook! " or "*Whoo*-ook!" Once in Paris, at the *pension* of Madame Melon-Hollard, in the rue d'Assas, my friend Dr. Rochat, from Switzerland, and I got into a talk, at luncheon, about cattle in his country and mine. After every-one else had left the table, we closed the dining room doors and compared the cow-calls that we had heard. (Hog-calling, cow-calling, and sheep-calling form a branch of comparative literature that I promise myself to explore, some day.) I discov-ered then, however, that this was a field of scholarship that re-quired literally a *field*. Madame Melon, her children, her staff, and half the pensionnaires came running in, white as sheets; I doubt whether the establishment has forgotten the episode to this day.

Once the cows are in, they pause at the watering trough

for a drink, and if the air is fresh and they feel good, for a bit of clowning. In the winter, sometimes you can break the ice in the trough with your feet, but sometimes you have to fetch a pick-ax, or a spud, or both, and hack away for a while. In the summer, the next step is merely to go to the house and fetch the milk pail and quart tin. Most farmers, including James, milk right into the pail, using both hands at once, and this is faster, of course, but I like to use a tin, because this makes it easier to keep the milk clean. My cows all know their proper stalls, and are all moderately gentle, even when fresh, so that I can sing to myself, and let my mind wander, and indulge in imaginary conversations.

Not all imaginary, either. Peggy's niece used to stand where I told her to, in a safe corner beneath a ladder leading to the haymow, and tell me about her sisters and playmates at home, and ask me all sorts of amusing questions about the cows and about everything else. I shan't soon forget that little girl, in a blue dress or sun-suit, matching her wide blue eyes, with pigtails hanging down behind. "Allen, is Eva Le Gallienne a sister of Anne's, or a cousin, like me and the twins? . . . Is Christopher Dudley's brother? . . . What's a half-brother? . . . They're hungry, *aren't* they? . . . Why don't you milk Douglas? . . . They're eating their supper, *aren't* they, Allen, just like me and my cereal."

The summer when Peggy was away, Jeanne, our little cocker bitch, never left my side, and always followed me to the barn. If she was late, and the door was shut, she'd jump in the window, and sit there, delicately, like some little lady who had happened into a strange environment. Whenever the other dogs come to the barnyard in winter, they get filthy, but Jeanne picks her way carefully from stone to stone, always watches where she steps or sits down, and comes back to the house as clean as ever. When both Peggy and I were away this winter,

James had the same companion in the barn. A few weeks ago there arrived at the barn a nice cat who had somehow lost one leg, but who is astonishingly agile, and quite useful as a ratter. She is a female, but we call her John Silver. On the whole Jeanne and John get along quite amicably.

Jeanne's cleanliness is more than I can claim for myself. When I come in with the milk, fix the strainer, pour the milk out, and get it somehow into a crowded icebox, there are two things that I need: a kiss, and a bath.

An invisible enemy.

Never before have I been so conscious of the importance of health, the possibilities of disease, and the necessity of hygiene. (The Pasteur movie, with Paul Muni, seemed to me an excellent piece of work, but it was probably for personal reasons that I found it the most exciting movie in years. When those sheep jumped to their feet, I could have stood on my chair and cheered.)

Yet I must confess that like most farmers, most men, I find it easy to forget cleanliness, and to rationalize my indifference by saying that I have done all that I can do, and by calling anything more than I can do easily, sheer fussiness. And like most intelligent women, Peggy has hygiene, in one form or another, almost constantly on her mind, and is capable of preferring worry and cleanliness to peace and dirt. If Peggy had her way, we'd think of nothing but bacteria, and be completely miserable. If I had my way, we'd never think of them, and be dead. Between us, we are still alive and happy. Without going into the arguments, dangers, and alternatives, which I leave the reader to imagine, I think I can report briefly our major precautions against disease.

Our old barn, a hundred yards from the pumps, has a gravel

floor, and the law quite rightly forbids us to sell milk. With some care, a liberal use of straw, lime, sheep dip, and fly spray, I can keep the old dump moderately clean. We wash the cows' udders only when they are noticeably filthy. Peggy keeps the icebox scrupulously clean, and every utensil that touches milk or butter is washed with boiling water and dried with special towels. When it is possible, they are all sunned.

Early in our career, at an expense of three hundred dollars, we had the cisterns largely rebuilt, with new filters and connections. Whenever we suspect the water, we have it tested by the Board of Health, and use distilled or boiled water in the interim. The cisterns need some alterations, and should probably be cleaned more frequently. This is difficult in a period of droughts and floods.

The small cellar, under the kitchen, has been a worry as long as I can remember. In a wet season, it is half or three-quarters full of water, and makes the whole house damp. When the house was treated for termites, the cellar was sprayed heavily with creosote; and it has also been cleaned and doused with lime. So far, we have been able to do nothing more about it.

The old privy had a mere hole, within fifty feet of one cistern. Lime, or disinfectant, or both, were dumped into it almost every day, and it was washed carefully once a week. In the summer, it was also sprayed for flies. We tacked netting all around the eaves and rubber all around the door. A few weeks ago, we found out that we could have a new and better one built by the W.P.A., in collaboration with the State Department of Health, by paying for materials only. We went ahead, and our new W.P.A. project is fairly splendid. It cost about twenty dollars. In addition to the Sears-Roebuck catalogue, it has reproductions of pictures by Toulouse Lautrec, Laurencin, and Chirico.

Another danger is an old quarry that we use for a dump. All garbage that can be eaten is fed; the rest, and refuse, are burned.

There remain tin cans, refuse that can't be fed or burned, and waste water. These are all put down the quarry, which is frequently limed.

The dog kennel is washed and disinfected frequently, and the hen house, when used, is scraped, limed, and disinfected about every fortnight. Evidently this is not enough.

A pile of manure rots in the barnyard all summer. We can't yet afford either a concrete pit or chemical treatment; we use dirt, but not enough. Yet we have no more flies than we had when the manure was all hauled out in the spring.

A man intimate with death.

Sooner or later, the best of precautions fail, and one is confronted with a question of cure, rather than prevention. A farmer has a good deal of nursing to do by himself. We try to do everything we can, but every now and then we have to call our veterinarian, Dr. Ames. One of this man's virtues is that he will tell me anything he can over a telephone, and is very glad to show me how to do anything that I can possibly learn how to do. I shan't soon forget a laboratory demonstration, so to speak, that Dr. Ames once gave me on the genito-urinary anatomy of cattle. I used to be fairly squeamish about blood, surgical waste, and giving pain in order to help an animal, but now familiarity, and the inherent interest of veterinary medicine, permit me to see and do almost anything, and lead me, before I know it, into boring and shocking people who haven't the same interest, and haven't had even my little experience. These things seem to me so marvelous, so real, so close, somehow, to what matters, that I lose all perspective and courtesy. I do think, though, that our Dr. Ames would seem real and interesting to almost anyone.

He is an old army man, who has served in various corners

of the earth, and kept his eyes open and his mind awake. He is an expert marksman, competes in national shoots, and provides instruction in small-arms firing. He has a son who is now an officer in the navy. Ordinarily, military men seem to me stupid and childish, but Dr. Ames is neither.

He is a huge, gaunt, powerful man, with stiff, dark hair, leathery skin, keen eyes, a happy smile, a gold tooth, and very strong yet gentle hands that at the right moments are clean. He is an ardent sportsman, and referees many of the local wrestling matches or football games. He is also the County Coroner, and takes care of all the people who have been hanged, or shot, or cut open by razors, by themselves or by a person or persons unknown. He is attached professionally to our favorite stockyard, and any Friday afternoon you can find him down there sticking enormous syringes of cholera serum or virus into the groins of screaming hogs.

He is always busy, but he always has time to talk, and is always worth listening to. Among other things, he is keenly interested in the prehistoric Mound Builders. He likes to read stories in the pulp-paper and slick-paper magazines, and to check up on the technical data of writers who enter fields where he finds himself at home. He is the only person in Ross County who is both interested in my writing and direct enough to give me advice about it. He tells me that I ought to make more exciting plots, study more queer and romantic characters, and try to write stories like those in the *Saturday Evening Post*. I feel that my talents, if any, lie in other directions, but I like to talk with a man who respects my trade as I respect his, and who can talk about it without any nonsense.

In fact, Dr. Ames has a very good idea of what he knows and what he doesn't know. There may be better veterinarians in the United States, but there are none better in this region, and I know that when, if ever, he finds himself out of his depth in some case on this place, he will say so, candidly. And I think it will be years, if ever, before that happens.

I don't know exactly why I find this man so interesting, so moving, even. It may be because in everything he does he is intensely alive, yet very intimate and casual with death. He could not have written, and he has never read, the things that have been said about death by those poets who have felt "the skull beneath the skin." His intimacy has a different quality; it is more crude and superficial. But I suspect that the fell sergeant is less afraid of poets, philosophers, and divines than of this old officer. . . . I don't know, but I do know that I am very glad to have Dr. Ames at the other end of a telephone.

A few clinical notes, expurgated, and easily skipped.

The worst disaster of this kind has been the death of Peggy's first cocker, Jerry. We went off to Chicago for a weekend. None of the dogs liked our colored people at the time, and we had, then, no dog-yard. We left Jerry and Jeanne on chains running on adjoining wires. The night we got home, exhausted, we found a note from the man, saying that the morning we had left, the dogs had got their chains tangled, and Jerry had been strangled to death. He was buried down on the east hillside, beneath an elm, beside my aunt's first collie. We have had this shame within us ever since.

After my aunt's death, Jack lay around in misery all day, and moaned beside my bed all night. Finally I wondered whether there might not be something else, besides grief: perhaps a chicken bone, or poison. Dr. Ames came out, and went over the dog carefully. He said it was nothing but grief, but that was serious. He said that if we didn't take the dog away somewhere, quickly, he might very well stop eating and die. I was still teaching, so my mother and uncle took care of him, for a year, on

Long Island. "Men have died from time to time, and worms have eaten them, but not for love."

Jeanne's pregnancy and whelping were surprisingly easy for her and for us. Dr. J. L. Leonard's book is good on this subject, as on others. Food, box, and after-cleaning are important. Docking doesn't hurt them much. Any puppy that is wrong, even slightly, must be killed immediately.

For dog lice, vinegar is no good, but "Tornado Powder" is. . . . Infections respond to alcohol, a dermatic solution, and alum powder. . . . Distemper and rabies shots are relatively expensive, but worth it, for peace of mind. . . . For bad ears, a few drops of a special, strong oil, or, lacking that, olive oil. . . . K.R.O. rat poison is an emetic only, but the meat used for bait can spoil and give real trouble to dogs. . . . Carrion, and too much milk stolen from hogs or chickens, can play hob. . . . Our dogs get wet in rain, and in swimming, but are never bathed. . . . De-ticking, in season, has to be quite regular. . . . Our dogs sleep outdoors, in unheated kennels, except when it is colder than ten below zero.

Chicken diseases are numerous, hard to diagnose without a laboratory, and practically impossible to cure. A postmortem and microscopic examination is a very neat job to watch. . . . Temperatures in the first few weeks are more important than food. . . . The only chicken I ever cured, and I can't remember now what the disease was, was stuffed with yeast. . . . Ducks and geese are tough nuts, but small ducks can go swimming in a watering trough, get stranded when the cattle drink down the water, and nearly drown when tired. But a few hours in the sunlight can bring around one that looks very far gone indeed. A guest, observing a duck in this condition, remarked: "'Too much water hast thou, poor Ophelia, And therefore I forbid my tears.'"

The cattle have needed less nursing than anything on the

place. Some years ago there was contagious abortion, which gives people, through milk, undulant fever, and my aunt got rid of it only by selling out almost entirely. Tuberculosis in cattle has been wiped out in this region, but the checks continue. Some years ago, one of our cows knocked the end off her hip bone in some way unknown to us. This gave her a permanent list, but did not interfere with her in any way. Self-cleaning, so to speak, after parturition, is essential. If it doesn't occur, your Dr. Ames can do it for the cow, in the most amazing manner. Any slip here can keep a cow from conceiving later, and nothing can be done about that. Sometimes cows strain themselves in winter mud, but ours have recovered in a few days. "Bag Balm" is good. Cows' milk cannot be used until three days after the birth of the calf. Before that, it will kill chickens and hogs. I should like to talk about castration and docking, and engage here in the debate about pincers vs. knives, and so on, but people less naïve than I am assure me that this subject is not really so interesting as I think it is.

Sheep have a most peculiar combination of delicacy and hardiness. They are subject to many diseases and accidents, yet they seem to survive and multiply under most adverse conditions. Almost all of them have stomach worms, many have diseases of the intestines, some have diseased feet, all have ticks in season, and many have trouble in lambing. I started well, because I got some good sheep, and none had been here before. I drench for worms, pine-tar for grub-in-the-head and colds, and take some precautions in lambing, but have not had to dip, until this year, when I shall use the watering trough, with a pint of "Black Leaf 40" to ninety gallons of water, or a creosote dip.

We have not yet had any trouble with dogs—and I'm pounding on wood. Sometimes at night I hear foxhounds and other dogs, and wake up in terror to hear my sheep bells, and the soft thunder of a hundred delicate hooves. I have a small pistol of Peggy's, but no shotgun. Dead sheep are paid for out of dog taxes, but that's money I don't want.

I think my interest in all this goes back rather far in time. I remember once when I was here as quite a small boy, and my aunt's best cow broke into green corn, ate her fill, and swelled up like a balloon. She was twelve years old, and had just given birth to a calf, and so it was serious. This grand old cow was the best in every way that we have ever had on the place, and was named after my mother. Dr. Ames came out, and my long career as nurse under his direction began. Every day I fed her enormous pills, crushed with a hammer, dissolved in water, and given from a long-necked bottle. The idea of opening her up was considered and rejected. After a week of this, she died. A man came out from the fertilizer works with a rickety old truck, and hauled the carcass aboard with a winch. He was a very fat man, and he had along a small boy who kept sucking a stick of candy. When they rattled away, the cow's hoofs, sticking out from under a tarpaulin, kept jouncing over the side. Both my aunt and I cried without shame.

A biological event not easy, safe, or wise to skip.

To the farmer as nurse, lambing time is one of the most arduous in the year, and to me it is one of the most exciting. I may have taken it so hard, a year ago last March, because it was my first experience with lambing, but from my feelings this second lambing time, and from the talk of other sheep men, I found that the excitement does not disappear, or even lessen.

A week or so before the time of that memorable first lambing, James and I clipped the wool off the ewes' udders and backsides, and began to pray for good weather. All the ewes had been bred, all right, and I had fed them well. Some looked much larger than others, and we tried to guess which would have twins, but it turned out that we had guessed wrong, in most cases.

The first ones, twins, came one fine afternoon, several days early, when Peggy and I were in town. James was on the job, and got them into a pen. As soon as Peggy and I got home, and had gaped and cheered at the funny little accordions in chinchilla, James and I got several other ewes that looked near, into other pens. (Incidentally, a ewe's time is 150 days; a cow, 283; mare, 340; sow, 114; bitch, 60; cat, 60; rats and mice, 22; canary, 13; chicken, 21; duck, 28; ostrich, 42. Nature seems to me remarkably fast. Figures from Sewall Wright.)

The purpose of making pens, and of keeping ewes and lambs in them, separately, for several days, is to protect the lambs and try to make sure that their mothers will own them. Often they don't, and a disowned lamb is a job and a half. A ewe tells her lambs by smelling their rumps, and refuses, in most cases, to let any other lambs nurse her. Sometimes they disown their own lambs, and to prevent this, it is wise to rub some mucous from the lamb's rump, immediately after birth, on its mother's nose. Kerosene on both is recommended, but didn't work for me. Sometimes they induce ewes who have lost lambs to take on other disowned ones by covering the latter with the skins of the dead lambs. Obviously, this instinct in ewes is to prevent general shuffling, which would probably result in all the lambs ganging up on the ewes with the most milk. However, the cause of the mistakes is less clear, and is as eagerly debated as the scent of foxes. My own little theory is that the quality of the mucous and its odor changes very rapidly on exposure to air, and must be detected by the ewe at once; also, that ewes can tell when a lamb has some serious defect, and is not likely to survive. The ewe acts throughout, not as the individual mother of an individual lamb, but as an embodiment of the flock's motherhood, caring for an embodiment of the flock's next generation. Only when a sheep is raised apart from the flock does it become an individual, with joys, sorrows, and mind of its own. Only when an animal has become humanized by long individual association with people, does it show any sentimentality—and not even then does it become sentimental about food or sex.*

We had only one disowned lamb, out of twenty-three, the first year. This one was the second of twins, and was born, to my surprise, at least two hours after the first. We had a supper party that night, I was not on the job for several hours, and when I returned to the pens, I found this second twin, disowned. Several times a day, for days, I held the ewe for this lamb, which we called Pip-Squeak, to nurse, and then I began with a bottle, and made sure that he got all he could drink. But Pip-Squeak seemed to get no nourishment from his food, and remained an accordion. In about a fortnight, he died. Another was premature, and a third was stepped on by cattle and paralyzed. Three lost out of twenty-three lambs from seventeen ewes was not bad, that year, for this region. Dr. Ames said I might never again be so lucky.

In animal midwifery, mother knows best, and when watched carefully, has a lot to teach. No help is better than fussing. However, I have a little kit ready, and take eight or ten unmentionable hygienic and dietary precautions that are in all the books but are not generally taken. They may be citified, but if they help, I don't care about that.

That first year, only one of my ewes had trouble. You are supposed to wait half an hour before helping. I waited three-quarters, disinfected myself, remembered the pictures in my book, and finished the process and the pain. This lamb became

* Moral-sociological note: I do not mean to imply that individualism is good or bad, or that sentimentality is always individual, or that emotional accuracy is always better than sentimentality!

the biggest of all. We called him Jimmy. I was quite alone with those sheep in the middle of a cold, raw night, with one flashlight. Along with moments of writing, my first trick alone at the wheel of a ship, my first class as a teacher, and a few others, that simple little act was one of the major excitements of my life so far.

It may be rightly inferred that lambing time is not a good time for parties, or for anything else. For about three weeks I could think of nothing else, do nothing else. During one week, when they were coming fastest, I didn't sleep for more than two hours at a time, and took off my clothes only to bathe, and to put on clean ones. Peggy, at the time, simply had to have some teeth pulled and spent that week in bed. I'm afraid I wasn't very good as company, or as a nurse, and I smelled, they say, to high heaven. I myself can smell almost nothing. Why, I don't know.

Our second lambing, this year, was, as I say, as exciting as the first. Perhaps because my new ram had been a virgin, the lambs did not appear on schedule, and I got worried, but they came on all right, and all within three weeks. We made the mistake of directing and producing a play at the Little Theater at the time, and Peggy bore the brunt of that; we both got rather ragged. But I had made better pens and other arrangements, and had more confidence.

I had my first loss of a ewe. She got down just a few days before she was due. Dr. Ames told me over the phone there was no hope for her; that other sheep men were having the same trouble, and the cause was not certainly known. But I wouldn't give her up. Peggy made a hammock for her of old bags; I kept her on her feet with this, and fed and watered her by hand, forcibly. After three days it was clear she would die. I clipped her myself, she died that night, and the next day we took her carcass to the free veterinary clinic at the State University, for a post-mortem. This I found fascinating. Twin ewe lambs, both too large, had been just due. They called it "pregnancy disease," and admitted its nature was not exactly known, but were sure it was not contagious, and thought it was caused by faulty metabolism. They approved of my feeding, but suggested more oats, and the sale of all my older ewes. Incidentally, I got what I thought was a good new metaphor to substitute for rats leaving a sinking ship: ticks leaving a dying sheep.

Meanwhile, the others were doing well. One, our favorite, had not been bred, perhaps because I had not clipped her backside at breeding time. The remaining fifteen had twenty-three good lambs, twelve bucks and eleven ewes. None of these was disowned, so far I have raised them all, and they are all eating grass and grain, which means that they are fairly safe, from now on.* They are not as uniform in size as I'd like, but most of them are well built and look like their blue-blooded father.

This year, two of the ewes had trouble. One discharged her placenta first, which baffled me and led us into a ridiculous search for lost lambs. But then I found they were coming, with trouble, and had to have help. They were big strong buck twins, and my relief was extreme. It has interested me to discover that after a successful and difficult delivery real obstetricians, however experienced, feel the same extraordinary relief and exhilaration. The other one came at night. Peggy and I had been painting scenery all day, and were tired. When we got home, just before James left, he told me one of the ewes needed help. He had tried unsuccessfully to reach me by telephone. I rushed out with lantern and simple equipment. She was in real trouble, all right. I thought the lamb was being strangled, and the uterus punctured. I worked as calmly and quickly as I could, in a cold

* Later bulletin: In July, five old ewes, a yearling ewe, and a ewe lamb, were killed by lighting that struck an elm tree under which they were standing. They were worth about forty dollars, and not insured. We skinned the lamb for a rug, cleaned the carcass, and took it to cold storage in town. We had been *too* lucky.

sweat of fear. Finally I got that lamb out, alive, without injury to the mother. I felt as though I had been at it for three hours, and had to sit down in the straw to rest and pant, with the mother. When I got back to the house Peggy told me I had been gone about fifteen minutes. I discussed these cases later with Dr. Ames, and he told me both ewes and their lambs would have died. Another ewe's milk was slow in coming, and her twins kept me up all night and the next day with bottles, but at last it came, and the lambs were not disowned.

Ten calves, including one pair of male twins, have been born on the place since I have been here, and with one exception they have been no trouble at all. The most exciting birth of a calf, so far, and for no special reason that I can see, was one that took place last winter. Peggy and I went to a rehearsal at the Little Theater, that evening, and drove home, about eleven, in a snowstorm. For the time being—and this is the chief virtue of the Little Theater, for us—we had forgotten all about our livestock, but on the way home we wondered about Anne, the expectant mother, bet the regular two bits on the sex of the calf, and guessed about the markings. As soon as I had put the car away, I got the lantern and went out to the barn. There was Anne, large-eyed and nervous, and there beneath an open window, with snow blowing in on him, was the new calf, not more than a half-hour old, and a beauty. I closed the window and went back to the house to get a pail of water for the mother, and a cloth of some sort to rub down the calf. I called the news up the back stairs to Peggy, who had undressed, but who threw on some clothes and came out to the barn with me, to see for herself. The little chap was still weak in the legs, but I got him to his feet and rubbed him down. We were glad to see another bull calf; more than glad, very much surprised, because he was the fifth in a row. Like all new calves, he was a little uncertain whether his first meal was at his mother's bow or stern, so I helped him to the connection, and we waited to make sure that he got some warm milk in his belly. Then we closed the door and came exulting back to bed.

Why exulting? We could claim no sort of credit for this, the most banal of nature's *tours de force*. This calf wasn't going to make our fortune; on the contrary, more than likely, sooner or later, he and his fellows would eat us out of house and home. I don't know. Raise livestock yourself, and find out. Look at any farmer, no matter how old, tough, and experienced, when he is taking care of his calves, or colts, or lambs, or pigs, or whatnot. Look at him, look at the small fry and their mothers, and feel the pulse of something beating. If odd things don't happen to your own circulation, stay in a penthouse, and be damned to you.

Axes and sun-spots.

Until I came out here to live, I was, like most city people, conscious of the weather only as a mild nuisance or excitant, as a furtive element in the streets, reminding one of other times and other places, and as the last resort of conversation.

Even now, Peggy is much more observant of it than I am. From many summers on the coast, and in sailboats, she acquired a permanent consciousness of the direction of the wind, and the promise of the sky. When her father is here, every morning before breakfast he taps the barometer, reads the temperature, looks at the sky in all quarters, and notices any change in the direction of the wind. Among seagoing folk like these, conversation about the weather means something, and I find it hard to keep up my end. James, of course, knows much more about the weather than I do. Still, I like to keep in my mind a general record of the weather in this one region, somewhat in relation to the disordered and precarious economy of agriculture, and in relation to the end (I hope) of three centuries of destruction of a continent.

In this region, during the War, the abnormal boom in commodity prices seems to have been nullified, in some small part, by extremes of heat and cold. Then, during the first part of the deep agricultural depression that began with rising world tariffs and increased production elsewhere, in 1919, ten years before New York and Washington discovered that anything was wrong in God's country, the weather seems to have been fairly moderate and helpful. Then, in 1930, there was a serious drought, that lasted, more or less steadily, until 1934 and 1935. The water level kept falling, and more and more wells went dry, while more and more banks closed their doors and more and more mortgages were foreclosed. In 1934 and 1935 we profited by the operations of the A.A.A., and, perhaps more effectively, by the disasters that ruined farmers farther west.

All during the spring, summer, and winter of 1935, rain fell almost steadily here, and we suffered from floods, not so serious as in New England and Pennsylvania, but important. The farm at Bellbridge did better than it had since 1919, although it got, from high water, a dose of Johnson Grass seed that became a formidable threat. For the first time in years, farmers began to look as though they might have at least stopped going down. Here at Oak Hill, we found ourselves living in a morass of mud, water, and steaming vegetation. The corn and the garden were practically choked out by weeds, in the summer, and in the autumn and winter harvesting the corn was a labor for Hercules, the livestock suffered from colds and straining in the mud, and our cars got bogged down almost daily.

There followed, 1935–1936, a winter more extreme than any since 1884. The temperature reached twenty-two below zero, and only ventured above zero furtively, now and then, during January and February. There was enough snow to keep the sheep from foraging, and to make driving up and down this hill a feat, but it was blown off, or melted, often enough to expose the winter wheat and rye to killing freezes. The fruit men lost all their peach trees, and most of their apple crop. Our rambler roses and grapevines were killed to the ground.

This broke suddenly, and there followed a month or so of delightful weather, in which everything got planted, and started growing with such promise that commodity prices looked shaky, and I sold half my corn at Bellbridge.

Then for about five months there was practically no rain, and extreme heat. For weeks and weeks the temperature hovered between ninety and a hundred in the shade, with occasional winds that felt as though they had come from a furnace. The wheat harvested in June was generally good, and the price was about a dollar. Old corn prices reached new heights, above wheat, but this didn't mean much, because most new corn crops hereabouts were about half of normal. Most gardens, including ours, which was our pride and joy in May, were wiped out. Most farmers were forced to feed most of their livestock in summer, or sell most of it, but although mine looked bad for a while, I held on without doing either. Pastures were browner and barer than they were in the following January, and there was almost no hay.

Towards the end of September, the temperature dropped suddenly, and there were occasional showers. There followed a winter of abnormal warmth and heavy rains, culminating in floods that—it can bear repetition—drove half a million people from their homes and did untold damage to buildings, crops, and farms. This spring, so far, has been mild and promising.

It should be remembered that compared to people in the Dust Bowl, and elsewhere, we have been fairly lucky, so far.

The Brookings Institution has reported what it was easy

enough to guess, that a very large number of the American people are underfed. If this is true, and I think it is, our problem of markets is an utterly false one, and both artificial crop reductions and efforts to stimulate international trade are, however useful temporarily, to small sections of the people, fantastic dodges of the main issue. I have already stated briefly the radical solution of this problem that seems to me the only reasonable one.

Many sound observers have also reported that one of the major causes of the droughts and floods of the last few years is the careless destruction, for temporary profit, of the forests and prairies: the ploughs and axes of our ancestors and of ourselves. The profit technique, which opened up this country to agriculture and industry, is now quite obviously ruining it for both. As an individual owner and operator, the farmer can do more than he has done, to conserve, and the Conservation Act is an intelligent effort to help him in this direction, while at the same time producing artificial scarcity. Essentially, it is as superficial as the A.A.A., the N.R.A., and the trade treaties. Mr. Kincaid and I cannot reforest the sources of Paint Creek, and build dams in it, and as our County Agent was intelligent enough to see, we could not, until the last floods, be expected to reduce further, for conservation of a superficial nature, incomes that are being wiped out by lack of conservation during the last hundred years, and are now dependent on artificially raised prices. We cannot, at one and the same time, atone for the sins of our forebears, try to feed the Germans, cut down on food needed in Columbus, Cincinnati, and Ross County by people who can't pay for it, try to raise purebred animals for people who can pay for them, pay taxes to the government, insurance premiums to mansions and museums in Hartford, and profits to hardware manufacturers and others—we simply cannot be expected to do all this at once and stay off relief. (If, by any remote chance, we are saved personally by my specialty, that will not affect the basic problem.) Here again, I can see only one reasonable solution.

Another cause of our extremes of weather seems to be the spots on the sun, which recur in cycles. Now I don't expect socialism, or any other human technique, to remove the spots on the sun, eliminate disease and death, and wipe away the tears from all eyes. Neither do I expect people in open boats in a storm at sea to still the waves, or walk them. But when I consider the markets, and the hunger, and all the rest, I feel annoyed and scornful. When I walk over good land and crops ruined by droughts and floods, when I feel the heat of the sun, and the cold from beyond the stars, I feel something a good deal larger than ourselves, something without any interest in us, something in whose presence we simply cannot go on being hogs, cowards, and fools, and survive.

Weather as a verb.

I cannot of course give any idea of our experience of the weather over a period of nearly three years. This great and indifferent power not only affects every decision we make, every moment of our work, and all its tangible results, to the last grain of corn, the last flower, and the last penny. It also, and this is probably even more important, affects directly the quality of every waking moment. In my case I have found that after fine weather of gentle rains and sunlight, which is both productive and comfortable, the next easiest to adjust to is the extreme, the freak, in weather. This is generally disastrous economically, and most uncomfortable, but it is novel, and gives one the exhilarating illusion of knowing the worst, of testing oneself to the limit, of coming to a hand-to-hand fight with the enemy.

The hardest weather, for me, is the weather of attrition, the last week of ten weeks of rain, the fourth flood, a dry, hot September after three earlier months of it in a row, the last short days of January, the last savage snowstorm in March, and so on.

As a sample of youthful pleasure in extremes, I quote a passage from a journal I kept for a while last winter:

"The other evening, the bottom dropped out of the thermometer, and the wind rose to a gale. I did all the chores myself, and sent James to the foot of the hill to rescue Peggy's car and take it to greater safety in town. When I got through, I came into the house with my heart pounding, and heaving for breath. A really cold, strong wind like this I find less disagreeable than the icy, penetrating dampness of Boston, for instance. But it does freeze your nostrils and mustache, paralyze your hands when you use tools, and finally knock the breath clean out of you, leaving you with a short headache. Peggy had planned to go to Cincinnati, that night, for some Girl Scout work, but she postponed it to the noon train the next day. We were fairly comfortable, close to the fire, and she beat me at chess the second time running. I went out to the barnyard at nine, and found all the animals intensely uncomfortable, but not in danger, despite the big cracks in some of the older buildings. I went out again at eleven, and by that time the thermometer had dropped to twenty-two below. Hunter, the young calf being weaned, alone in his little shed, was shuddering with cold. The others, and the sheep, still seemed safe. I began some shuffling, but only succeeded in getting a steer in with Hunter, to share body heat. Instead, the steer kept butting the calf, and chasing him around, which was just as effective. With the help of Jack, who obeyed me perfectly this time, as he generally does in a pinch, I chased the whole herd around through the drifts for a while to warm them up. Finally I went back to Hunter, to give him a last rubbing up the spine. I had propped the top of the rumble seat, from my old Ford, against the door of his shed, outside, to keep it shut, and so I locked myself in. It reminded me of the time when I was pitched into a locker in the chart room of the *Santa Clara*, and the door slammed behind me. Shoving, jiggling, and laughing, I managed to get out. I thought I was probably being fussy and foolish, until I learned that a neighbor at Bellbridge lost six calves that night, frozen to death.

"It was really too cold in the kennel, so we kept the dogs in the house, downstairs. They got cold, and raised cain all night, rolling over furniture, scratching doors, whining, howling, and barking. Peggy and I kept up an intermittent debate: Would it be a good idea to beat them up? If so, who would get up and do the beating? Who could beat whose dog how hard? Damned silly, but who wouldn't be, under those circumstances? It ended with Jack's turning a door-knob with his teeth, and sneaking upstairs, and my chasing him down and beating them all up. This was unpleasant, and only mildly effective, and we were all fairly exhausted the next morning.

"The weather had warmed up to twenty below. James got here half an hour late, I grabbed him, and we did the chores in practically ten seconds flat. The cattle looked done in, and a couple of sheep had running noses, but everyone was alive. At eleven Peggy and I tramped down the hill with a suitcase and a change of galoshes for her. We got her car started all right, but I backed it into a ditch, and she had to drive it out. We lunched in town, and supplied with a *Harper's Bazaar* to help her preserve the illusion that she was still in a position to read that magazine for practical instead of fantastic purposes, she caught her train. I went to my old colored barber for my monthly haircut. The King had just died, and they kept asking me how much an English king could do, but I was too tired even to be interested in their interest, or to think up any answers. Then I came home and helped James with shoveling. After chores, I cooked myself a solitary and miserable supper, but then I got so excited about an argument, by mail, about Stakhanovism, that I let the fires go out, and the house, including my bed, was like a cold storage vault. I took three dogs to bed with me. They needed my heat, and I needed theirs.

"The next morning it was even warmer: fifteen below. The orange juice froze before I could get it squeezed out, which seemed fantastic, but we are used to that now. Also, to having water freeze in glasses on the table, and to cooking in mittens and muffler. I got Harold Prentice, our garage man, to get both cars going, and we finally got the road open for a few hours. Then Harold and I went to a little bar in town and had a couple of drinks with beer chasers. He had pulled out thirty-eight cars the day before and had some crazy yarns about them. A neighbor of mine came in and we all got very agricultural. Also, slightly tight. But I got away and got my shopping done. . . .

"The next morning it was two above, and everything had eased considerably."

But here are a few more notes, made several weeks later:

"Night before last, and yesterday morning, it was twelve below again, and yesterday was grim, so that finally we both got depressed. One can fight this cold with some pleasure for a few days, but after a while one's resistance goes, whether one likes it or not. It is impossible to do the littlest job, outdoors or in, without a surprising and conscious effort. Fingers and mind go numb, and spirits go down, and down. All we can do is huddle about a fire, trying unsuccessfully to read, or knit. My morning and evening chores are a relief, because they rouse me, but other jobs outdoors are practically impossible. We even find it hard to think of things for James and ourselves to do. Everything seems a huge and silly effort. Last night, milking, I tried to cheer myself with an imaginary conversation, and failed. When I came in, we had a couple of drinks, read our mail again, and felt better. Once more we debated, and rejected, the idea of flight. For escape literature, we read aloud from Dorothy Sayers, W. B. Yeats, and W. Shakspere."

How to undress a sheep.

Finally, in spite of everything, and to one's surprise and relief, there are a few results. Nothing startling, you understand, nothing to discourage the Administration about crop reduction, and nothing to encourage the merchants with whom one trades, but something. The fine old word harvest is rarely used: wool is clipped or sheared; hay is cut and made; wheat is cut or reaped and threshed; corn is cut and husked, or shucked. Still, I am going to use the old word anyway.

Except for early lettuce, radishes, and asparagus from the garden, the first harvest of the year is the clip of wool from the sheep. The shearing is done as soon as there seems to be some slight chance of no more really cold weather, and as soon as the sheep begin to be uncomfortable. By that time, the wool is nearly three inches long, some yolk or grease has risen in it, and it is not yet fouled by burrs and tags of manure. Our first sheep year, the lambs came in March, and we sheared towards the end of April. We were earlier than most of our neighbors, but the wool was in good condition. One of my ewes had begun to lose her wool; we saved what we could, and included it in her fleece, and I ought to have sold her that fall, but I didn't. This year we sheared at the same time, and the same ewe has begun to lose her wool.

The slight trimming and cleaning that I had done before lambing, that first year, had taken so long that I knew that I could not do the actual shearing myself. I was a little ashamed of this until I discovered that most men who have raised sheep all their lives can't shear their own sheep, and have to call on outsiders. I had heard so many tales of the carelessness of professionals, who often, it seems, nick the sheep badly, that I thought of watching a neighbor shear, and then borrowing his power shears and going to it myself. But I don't like to keep borrowing tools, I didn't want to botch the job myself, and Dr. Ames told me of a man who he said was very careful and efficient. We located this man, and a date was arranged. He had warned us to keep the sheep dry, so I penned them up the night before. From the top of an old table, I made a shearing platform about eighteen inches high, and six feet square. It should have been a little larger, but it sufficed. That morning I got up early enough to get all the milking and chores done before seven-thirty, when James and the shearer would arrive. I had left the upper doors of the shed open, for ventilation, but it had rained, and some of the rain had blown in on the sheep.

The shearer arrived on the nail. He was a tall, thin, powerful young man with red hair, and he brought hand shears, which he preferred. He felt the sheep, and at first decided that they were too wet to go ahead, but soon discovered that only one had been really wet, and that we could dry her fleece on the floor of the corn crib after she was clipped. We put the platform near the doors, and the shearer set to work. I caught the sheep for him, weighed the fleeces, and put them in a huge bag I had got the day before from the Farm Bureau.

Shearing is an extremely nice job. The wool has to be clipped close to the skin, which is loose, tender, and not sharply different from the base of the wool in color. There is no shearer in the world who can work without any nicking at all. I was nervous at first, but quickly reassured. It was apparent that my shearer knew his job thoroughly. When he finished, there were not more than a dozen nicks on seventeen sheep, and only one bad one. The yolk is itself a disinfectant, and most farmers don't bother with anything else, but I had ready a weak phenol solution, which my shearer was glad to use, although he pointed out that cuts and scratches on his own hands were always healed by the grease from the wool. I thought I'd make a fortune out of this idea until I discovered, with no great surprise, that lanolin, refined wool grease, had been used as a base for ointments for many years. I have heard that an isolated man in Minnesota or somewhere "discovered" the principle of the screw!

The shearer sits the sheep on her rump on the platform, leaning against his own body, and held by his left knee and upper arms. Then he begins to clip beneath the right ear (or left ear, if he is left-handed) and works down and out, taking the fleece off in one piece, just like unbuttoning and slipping off an overcoat. As in all work with animals, you have to be fast and sure: while clipping close and fast, without nicking, you have to keep the sheep from struggling and keep the fleece from falling apart and getting mussed up. It is one of the neatest little feats of manual dexterity and skill I have ever seen. For my seventeen sheep, that year, my shearer took about two hours and a half, which I think is very fast, considering that all the time we were talking about sheep, horses, and hogs.

The father and grandfather of my shearer had been shearers too, and he was using their very whetstone. He sheared between twenty-five hundred and three thousand sheep a year. He was planning to compete as a shearer at the state fair, but I did not find him there and was disappointed in being unable to cheer him on. He was also a traveling farrier and blacksmith, and raised hogs from garbage he got at restaurants. He had the shrewdness, toughness, good humor, and insouciance of all good workmen and free spirits, whether they are actors or plumbers, writers or house-painters, composers or veterinarians.

After a ewe is entirely shorn, the fleece is gathered together, with the clean inner side outside, and tied in a ball with paper twine. Everyone has been told of the necessity of using paper twine only, but when I visited a worsted mill and saw the labor and expense required to remove bits of fiber from other kinds of twine, I got the point. Manufacturers of woolens might save money by getting together and equipping and staffing railroad car or bus exhibits, in miniature, of the processes of grading, scouring, carding, and spinning wool. In tying a fleece, you have to be careful to keep from mussing it up or breaking it apart. This is one job that I did learn how to do.

As the job progressed, there was more and more bleating in the shed, and outside, because the lambs could not recognize, at first, these new, thin, clean creatures, their shorn mothers, but the mothers could recognize their proper lambs, as always, by smelling their rumps. There was some fighting between ewes, who did not recognize each other. They say bucks that are old friends will fight savagely after being shorn.

As soon as the job was done, I paid the shearer his twenty cents a fleece, loaded the huge sack of wool, with a crate of old hens I had decided to sell, onto my little truck, and drove to town. (I shall explain shortly the method of selling wool through a Coöperative.) This year I had two large sacks. Farming has its low moments, but this act of delivering the goods that people need is not one of them.

Not for sale.

The next harvest, still excepting garden, is that of hay, and for reasons that I have already, I hope, made clear, it is our most important harvest here at Oak Hill.

This last year we had a little clover to cut in the thirteen-acre West Riding. It was very bad in the part that had been marshy the year before, not even worth bringing in. On the better drained slopes it had felt the drought already, and had some white top in it, but was definitely worth making. Clover is cut just after the blossoms have begun to turn brown. As always, I had to spend most of a week driving around looking for a team and tools. Finally my very good neighbor Ralph Stone, at real inconvenience to himself, sent me a man with a team and a mowing machine. He also loaned me a small hay wagon and a wooden rake, which James and I fetched with the truck.

To save expense, to keep Stone's man as short a time as possible, and to get the hay in quickly, James and I used the wagon, with the truck as a team, while the other man mowed and raked ahead of us. My little old truck made a funny-looking team, but it did the work. Hay has to be dried by the sun enough to prevent its moulding later, but not so much as to let the leaves drop off and most of the food value evaporate. You can tell the right time by taking a handful and twisting it: it has a certain feel, a dry resilience, that Mr. Kincaid showed me. It was a very poor crop last year, only two-and-a-half or three tons, which we ricked with some I had bought, leaving my tiny hay mow for better purchases. This year we got about five tons on the place.

But I like making hay, no matter how small and poor the crop is. Even I can smell it, the hot sun feels good to my very marrow and soul, I like the violent exercise, with sweat streaming from every pore, and I like to think of how good even those few tons will look in the winter. It is fun to work like that, with a good man who doesn't talk except when he has something to say, and with all the dogs playing around and resting in the shade beneath the wagon when it stops. While we were finishing, the man with the team and mower clipped the remnants of rye that the cattle and sheep had not eaten in the East Riding, so that everything was clipped, and every wisp of grass was cleaned up, and the sod looked like clean yellow-green linen, fitted tightly to a beautiful body. The last night, a good rain came down, soaking the roots and giving the grass a fresh start.

That was the last good rain for months. The old grass in

one field, and the new in the other, and the pasture, and the garden, and the lawn, and the very leaves on the trees turned brown, old, and thin. This spring I begged another team and tools, bought some more hay seed, and started all over again. You can't lose every time, and I reckon that memories of sunlight soaking into the inner places are safer than any deposits in any bank.

The body and the mind.

A few days after it has been drilled into the earth, in the autumn, the wheat appears, bright green beneath, as it seems, the old corn shocks. Then in the winter it grows dull, but in the spring it turns green again, and fairly shoots from the ground. The green leaves put up stalks of green straw, with green heads that look (and may or may not be) heavy with grain. The old corn shocks have at last been hauled away, and there are miles and miles of soft green wheat, usually free of weeds and caressed by the wind, like the sea.

Then one day you notice that the green has paled. A few more days, and it begins to turn gold, and before you know it, the binders are in it, like huge insects that do not eat, the sheaves are piled carefully into shocks, and the wheat fields look like filigreed gold, with green velvet showing through, where the corn shocks have been.

We never have wheat here at Oak Hill, but I drove a binder a few rounds once as a boy, and that machine still seems to me a wonder comparable to linotype or monotype. It cuts the straw, lets it fall flat and even, gathers it together into sheaves, wraps these with twine, ties the twine with a knot, and ejects the sheaves at the side. Out West, of course, with their huge fields and combines, they thresh at the same time. Small combines are now invading this region, and it may be that, in a few years, threshing day will be an event remembered only by old folks. I for one shall regret this, because threshing is a grand scene, and also a symbol and lesson in coöperative action.

A few years ago, in this region, most of the threshing machines, and the steam engines with which they were powered, belonged to men, not farmers, of the mechanic type, who toured the countryside with them. Nowadays, more of the larger farmers own threshing machines of their own, drive them often with tractors rather than steam engines, and thresh only for themselves and half a dozen neighbors. Until last year, Mr. Kincaid did this, but now he has traded his separator and steam engine for a tractor, hires his threshing done, and for all I know, may have been—until these floods—wondering about these new little combines.

Without a combine, ten to twenty field hands are required, with teams and wagons, to pitch the wheat sheaves from the shocks onto the wagons, and haul them to the threshing machine. And so for a week to a fortnight, in early July, most farmers are busy in their own and their neighbors' fields. In this way, the work gets done with little or no money passed from hand to hand, and if there's anything a good farmer likes, it's a trick or technique that prevents that. At the same time, farmers' wives and daughters are busy preparing gargantuan meals to feed the influx of men, upon their own farms, on the Big Day. The machine often arrives the evening before, and very early the next day, the steam whistle on the engine announces to the men who already have their wagons loaded that the actual threshing is under way.

Pitching wheat is good fun, but not easy. One summer when I was about fifteen years old, I decided that I wanted to work in the wheat fields, and so when the threshing machine appeared in the valley to the west, I went down and asked our neighbor, Mr. Sam Stone (Mr. Ralph Stone's uncle), for a job.

He was amused, and told me I could carry the water jug around to the men. But this was a small boy's job, quite beneath me, I felt, and I told him I wanted to pitch. All right, he said, and assigned me to a wagon. By noon, I was almost in a state of collapse, and when I sat down at the huge and groaning table, I couldn't eat enough to avoid remark. Everyone thought I was sick, and wanted me to go home, but I denied this, refused, and finished the day. As a matter of fact I went on from farm to farm for several days. Not long ago, I bought some bean and tomato poles from Mr. Sam Stone, and he recalled that day. "I never thought you'd get up that hill that night," he said. "I came near tellin' one of the boys to hitch up and drive you home!" I'm glad I never knew how near I came to ultimate humiliation.

A few years ago, I was bicycling in France, at threshing time, and one day about noon, somewhere in Picardy, I stopped at a barnyard where one of their diminutive machines was set up and working. There were only about ten hands, and half of these were women. When they stopped for lunch, they merely sat down in the straw, and the children fetched loaves of bread, hunks of cheese, and bottles of wine. I talked with them, as well as I could; they were curious and friendly, and invited me to share their lunch. I lied and said I had had lunch, because there seemed hardly enough to go around as it was. Beneath superficial differences, those men and women, like the men I watched dipping sheep near Nether Stowey, were very much like my friends and neighbors here. And obscure scholars whom I have met in Europe are not unlike obscure scholars I have known in this country. And so on. (As for those who confuse and loosen the bonds made between men by the work that they do, to feed their own bodies and minds, and the bodies and minds of others, may they burn and rot in living hell, this side the grave.)

But more exciting than any of this, in the harvesting of wheat, is the wheat itself, waving golden in the fields, or resting in sheaves and shocks, or heaved up by brown, dusty men and going through the machines, or running in thin streams into the sacks. There it is, food for body and mind, body and mind itself. It is very close to the ultimate mysteries. This has been felt, and almost said, by E. A. Robinson, in a sonnet called "The Sheaves." This harvest is like sowing seed, or the births of the animals, or the making of wine. Flesh and blood, bread and wine, seeds and death: no thank you, gentlemen, you can keep your offices and trains, your files and accounts, your wing collars and umbrellas.

Thanks to a playful angel.

Ever since Eve got ambitious for herself and her husband, we have all had to earn our livings by the sweat of our brows, but even if St. Michael did stand over our first parents with a flaming sword, and even if our factories do stand over us with flaming furnaces and filthy chimneys—or locked gates—even so, there was, I think, a sweet and humorous and playful angel, not in the best of favor in Heaven, who sneaked out of Eden with Adam and Eve, and afterwards, when they got tired of their hoes and accounts, led them to fishing streams and to abandoned orchards, to old pastures with mushrooms, and to blackberry patches.

Every wet year we have tons of big, juicy blackberries, and although all the patches are owned by someone or other, and can be good-humoredly reserved in part, it is generally admitted that anyone can have all the blackberries he can pick. So every good summer Peggy and I and everyone else in the county sally forth with buckets and get delicious desserts, with rich cream, for supper, and enough to can into the most marvelous jam for breakfasts throughout the year.

But this angel was a joker, and he invented jiggers. The

first summer I was living here, I forgot how bad these could be, and spent a day in the thickets on the east hillside, with nothing on but moccasins, pants, and shirt. Well, that night my wrists and armpits, and every square inch of my body from waist to toes swelled up into enormous, horribly itching welts, and I didn't recover for a week. The only help, and it isn't a cure, is chloroform. As preventives, thick brine, or coal oil, smeared all over the body before berry-picking, have been recommended to me, but brine is not very effective and coal oil is not very attractive. The best thing, I have found, is a very heavy dusting of sulfur, all over the body. If you do that, wear tight clothes, and tie up your wrists and ankles, you may get away with it.

And it's worth it, not only for the berries, but also for the fun of picking them. You plunge into remote thickets, far from everyone, and your dogs hunt in the bushes, invisibly but audibly, all around you, and you lose any extra pounds or ounces in sweat, and your legs and arms and back and front and scalp are scratched red, and you come home with a heavy bucketful of good food that you feel, somehow, you have stolen right out from under all the solemn iron laws and systems.

Mushrooms, of course, are much easier, if you are up betimes, and in my opinion, even more delicious. Here the joke —and appropriately, perhaps—is more sinister. We happen to have forgotten most of what little we ever knew about the different kinds, and have not gotten around to studying them, so we stick to two kinds that we know. We have sometimes wondered about raising mushrooms, but think that the capital outlay, to get the right temperature and humidity, is beyond us.

Suggestion for unemployed condottieri.

But the real victory is in the harvest of the corn. It is on these enormous fields, drilled with such precision, kept clean with such care, with the green pennons and tassels waving all summer in the wind, and harvested with such a fight, that all of us, in this region and elsewhere, eat and live. It is these heavy ears that keep the big parade of hogs and cattle moving from pasture and barnyard to stockyards, slaughter houses, kitchens, and dining room tables. It is these yellow seeds, put carefully into good and bad ground in May, that become, in time, not only corn flakes and bacon, but also figures in checkbooks, permanent waves in hair, curtains rising in darkened theaters, and even strange ideas and emotions set down, curiously, in black and white marks on paper, and moving strangely into people's heads, like toxins and antitoxins.

When I moved out here, the meadows had to be plowed up, and so for two years even remote, uneconomic Oak Hill heard the portentous whispering of the corn in the wind. In August, depending on the weather, the stalks and leaves begin to turn yellow, at the bottom first, and the ears, if the crop is good, push upwards, outwards, and downwards towards the earth. Sometimes in midsummer we have the storms and high winds that destroyed the *Shenandoah* and her men, and that flatten the corn to the ground, so that it does not ripen properly, and is very hard to cut.

Along in September, we begin to see whether the grains are dented, and look to our corn-knives. These are formidable weapons, with straight, two-foot blades, blunt at the end. With one good blow they will cut a hill of corn (or a head, for all I know) clean off.

First, you find the four central hills in an area fourteen hills square, bend down the tops, and tie them all together. This forms a "gallus," not to be cut until the fodder is hauled in, that is the core of the shock, and provides something against which to lean the first armfuls of fodder. Then you begin to cut, hacking away with your right arm, and gathering the tall stalks, heavy with their ears, into your left arm, until you have cut so many that you can hardly carry the load. One soon learns to balance that heavy load on the hip, so that the entire weight is not on the left arm and shoulder. It is well to tie the left cuff to the thumb, or to a glove; otherwise a few hours of cutting will rub the wrist raw. The armfuls of fodder with corn are piled into the shock, and it is very strange to me that the sound of a load moving through the corn not yet cut, and stacked against the shock, makes a sound exactly, but *exactly*, like that of a wave breaking on the shore. In corn-cutting, my mind is always full of memories of the sea. When a shock is finished, you take a thing called a fodder pulley, which is merely a notched board fastened to the end of a light rope some fifteen feet long, throw it around the shock, and pull it together as tightly as you can. When you have got the pulley tight, you tie the shock together with binder twine, and then take off the pulley. Then you go on, for hour after hour, and day after day, until finally your field is no longer a West Point parade, but an even array of fortresses, with ripening pumpkins in between.

Corn-cutting, like most farm work, is hard enough. I remember well the second day of my first corn-cutting; the first day is always deceptively easy. About ten-thirty, the field began to slant and rotate. I had to sit down in the shade for a few minutes, and then with lunch, or almost instead of lunch, I had to take a stiff drink of whiskey. (In a pinch, there is nothing like it.) But corn-cutting is a good job, a constructive, satisfying release of all the sadism frustrated and stored up for a year, or years. The air is fine, with the smokiness and slight chill of early autumn, and the work itself makes one feel like Alan Breck, fighting all day with a cutlass.

Cutting is only the first step. Next comes husking, which can be postponed almost indefinitely, and stretched out through autumn and early winter, but which has to be done, sooner or later. It is rather simple, and rather dull. You untie the shock, push it over, and then rip the ears out of the husks, and throw them into a pile. Then you put the fodder back as it was, and tie it together again, to hold it from the wind until you can get it hauled in, and stacked near the barnyard, for feeding. There is a little hook, mounted on leather, that you can get and tie onto your right palm, and that makes it much easier to jerk back the husks. If cutting would be good for fighters, husking would appeal to the acquisitive. Those little piles of yellow beside the shocks, gotten there with so much effort, are literally gold. They may cost you more than they are worth, but that consideration has never yet deterred the counters.

The next thing is hauling in the fodder and corn. This job, too, can wait almost but not quite indefinitely. The winter before last was so bad that one saw piles of corn, and shocks of fodder, in the wheat fields as late as spring. Some good farmers, like Mr. Kincaid, haul their corn in right away, and their fodder as they need it, or as soon as they can. If you are in the bottoms, and don't get it in right quick, there is always a chance of its going down in a flood and landing in someone's bedroom in Memphis. If you have teams and wagons, only the worst mud can hold you back, but if you are trying to farm with an old roadster turned into a pick-up, it is another story.

Last winter, James and I labored with that corn, that fodder, and that damned (but invaluable) truck, all winter. We'd wait, apprehensively, for a hard freeze without much snow— the worst possible weather for the rye or wheat coming on— and then rush out, drive and load the truck as fast as possible, and sooner or later, get stuck. Like most city people, I had the idea that when your car got stuck, all you could do was get a team or wrecking car, but soon enough I learned what can be done with fodder, sacks, rocks, old boards, old fence posts, and the most grueling work. Several times I got Peggy into it, to drive while James and I lifted and pushed, finally got her moving, and then waited, with our hearts pounding, while Peggy drove with mad skill for the barnyard. Usually we had to un-load all corn and all fodder first. James is made of iron and steel, but more than once I'd come in, plastered with mud, and chilled to the marrow, from a three-hour session of this kind, and have to lie flat on my back on the floor by the fire for a couple of hours, praising God for the Jameses of this world, and loathing my own weakness.

But we got all that corn into the crib, and all that fodder into the barnyard, and both into the bellies of the sheep and cattle, and the manure back on the earth, and there are people somewhere who had woolen socks on their feet, and ate roast beef for dinner. And spring came at last, and Peggy and I, and James, put more seeds into the earth.

Repairs and laughter.

When we were in college, a friend of mine and I worked for a couple of months as "deck cadets" on a small steam freighter. One of the many things we learned was that intellectuals don't need to fear manual work, and another was that they don't need to feel inferior to manual workmen. We may be weaker, physically, and we may be less experienced, but we are generally smarter, and we have, or ought to have, better internal equip-ment for enjoying the work.

Another thing we learned was that if you have a hammer, a chisel, a pair of pincers, some nails, staples, wire, old lumber, a bit of paint, and a brush, you can fix almost anything well enough to last a while. I'm not saying I can do it; I'm saying it can be done, by good sailors and good farmers. On a farm, as on a ship, it's a good idea not to throw anything away, and it's not a good idea to rush to the nearest hardware store. I often think admiringly of the sailors I knew, and of the boys and men in Masefield and Conrad; of the farmers I know, and of Robinson

Crusoe and Adam. In forty years, I may be handy about the place.

Any old house and farm, even when they're in good repair, demand almost constant repairing and doing over. Peggy has done an unbelievable amount of scraping, painting, tinkering, and rearranging, and I have done some, but a good deal less. In the winter James is essential for a few things that we can't do alone, and a luxury for other things, like dishwashing and chores, that we can do, but even so, there are times, when outdoor work is impossible, when we have to set our minds to finding things for him to do. The answer is usually painting and repairs. With the preliminary help of a hired sanding machine, he has done over three floors, quite admirably. He has also done over some furniture, including a $75 corner cupboard that Peggy bought at a sale for $7.50. Except for rebuilding, and house-painting, he and I have been able to do the farm repairing, but for those matters, and for certain finer little jobs, indoors, we have had to call now and then on our old Negro carpenter, and painter, Mr. Sligo. This Mr. Sligo is an expert workman, a gentleman, and a character. He has a partner named Bill, and the two of them converse with animation and gaiety all day long. I am ashamed that I haven't yet learned how to do all our repairing myself, but if I had been a perfect handy-man from the start, I never should have known some very interesting, amusing, and valuable people.

The only kind of repairing I have learned something about is that of fences. This is natural, because fences are the *sine qua non* of animal husbandry; they are the only means of controlling feeding and sex; they are the very symbols of domestication. And they are always getting out of repair.

In this region, we use yellow locust for posts. A good post of this kind will last twenty years. We have some yellow locust on our west hillside, and James and I have cut and split only a couple of dozen posts so far. We haven't as many mature locust trees as we'd like, and those we have are on a cliff. Cutting and splitting are fun, but dragging posts up a cliff is another matter. We have used a few iron line-posts. They cost forty cents apiece, but are sometimes worth it, if you have it. Fence-building and repairing is a nice craft, because the stresses and strains are greater than one would imagine, and they don't all occur in the obvious places. There are several schools of thought about the bracing of end-posts, and one good fence-builder I know refuses to set posts in the dark of the moon.

Last autumn I had a scare that will, I think, keep me conscious of old fences and their repair for many years. Eight ewe lambs knocked a board off a fence and spent the night with our new ram. It was a chilly night, and it was just possible that some or all of them had "come in." Now a ewe should never be bred until her second year, and I was quick in calculating that if the ram had got them, they would lamb in February. And a ram is rarely shy and hesitant. Once I heard of a ram that got through a fence one afternoon in August, for three hours only, and the following January, forty-two ewes dropped lambs. Well, it is now spring, and my ewe lambs were *not* bred, but a good fence is more reliable than luck.

It always seems to me that there are a good many major words, and major questions, implicit in a posthole digger, or a pair of pincers, or a can of paint.

For instance, good fences go, so to speak, with good land, buildings, tools, livestock, food, water, house, and black figures in account books. They are like getting up early every day in the year, never running anything down, never forgetting anything, and never spending any money for fun. All these things mean good farming, and if a farmer isn't a good farmer, what good is he? Often enough my ignorance, carelessness, neglect, and half-amateurishness, when vividly demonstrated, send me into the depths of depression. As you will soon see, we spend a good deal of time, and energy, and even a certain amount of money, on reading, drawing, play-making, and other activities that are not often associated with farming, partly because they can interfere very seriously with it. Often when I think of these goings-on, and of some of my boners as a farmer, I feel like a bad poet with a velvet jacket, a lily, and dirty underwear.

But isn't good farming a means? And if so, can it be so emphasized, refined, and perfected that it becomes a thing of beauty, and an end, in itself? Once at the Winter Circus in Paris I saw a juggler so extreme, so imaginative, so exquisite, so perfect, in his ridiculous job, that he demonstrated to me that any means whose end is not destructive to life can be exalted to an end in itself. This juggler died a week later, and it seemed to me that we needed a superlative funeral for this man more than for most statesmen. But that silvery, that ultimate perfection is so rare, in any activity, and costs so much, that I imagine most of us do well to think a little about ends. I am not sure that I can imagine a farmer, and his farm, so perfect, as such, as to justify the distortion and misery of any man. Good farming is more complex, even, I think, than good juggling or good painting, and the odds against perfection are even greater. I have never seen a farm that justified a sour face, and too many good farmers go around looking as though they were attending their own funerals, as indeed they are.

The ends of farming are certainly not natural; that is to say, nature has never made a farm apart from man, as she makes some deep-sea fishes, just for the hell of it. We are constantly fighting her, and persuading her to be our ally. To this end, we constantly cheat her, and play tricks on her, and often enough she drowns us or buries us in sand for our shortsighted pains. No, if there isn't a good human excuse for it, there is none. The economic and social end, to feed ourselves and society, is obvious enough. But we can feed ourselves, as individuals, without repairing so many fences, keeping so many books, and wearing such long faces. And as for society, if it is so stupid as to half-starve itself, while hiring us to grow less food, surely our feeding *it* isn't so important as to justify our distortion, the clouding of our days.

No, I can't escape the conclusion that the quality of a farm, or of a job of farming, is precisely the quality of the farmer. We are only sensible to feed ourselves first, if we can, and that means feeding our senses, our minds, and our emotions quite as much as feeding our stomachs. At best, a farm is a very inefficient food-making machine, but I think it can be a good man-making machine. Repairs and improvements are always doubtful in outcome, but a color, or a shape, or a dream, or an emotion, or an idea, or a smile, is always beyond question.

So it is that I prefer those farmers whose farms may or may not be on the messy and mortgage side, but who always smile and wave Hello, and who often have time to stop, watch, and wonder, alone or with a neighbor.

And so it is that when Peggy gives herself backaches, and asks me to make the fifth repair in a row, she may encounter my indolence, unhandiness, and fatigue, which give her a cold shoulder, but she is just as apt to encounter this idea, and a feeling, which together give her a special kind of laughter.

Outbuildings and the Zeitgeist.

Effective repairs are always better than tearing down and building anew, but there comes a time when the most clever repairs are no longer effective. Even then, a surprising amount of the old materials can be saved and used again, either because they are still sound or simply because they are nice to look at. Now at last, however, our Nordic prudence, sentimentality, and fogginess can be put aside, and we can let ourselves go, in a more Latin mood: we can tear down, clean up, and try to fit the idea to the need, and the building to the idea.

As a tool for living, that is, for keeping warm and clean, and for eating, drinking, bathing, sleeping, working, and entertaining, our old house is very badly designed, and inefficient. However, our delight in it suggests that there is much more in living than in these more obvious activities, and that in all the others, our house is very efficient indeed. We have put in electricity, and Peggy has made countless cheap improvements, but we have not been able to afford any others. If and when we can afford them, we shall be most careful to keep them out of sight, and try to get gadgets that don't require constant servicing. This house presupposes more money and servants than we have, but it looks like a farmhouse, is right for the landscape, and is beautifully proportioned. It looks like a house for people with more dignity, substance, and children than Peggy and I can claim, but oddly enough, it feels sympathetic to the new things, to the restlessness, to the extremes, to the vagaries, to the experiments, and to the hazards that come from within us. It has seen a number of rather different generations, and now it welcomes us, and ours. Also—and this may be more important than any of the rest—it looks and feels somehow as though, barring fire, it could survive with grace, dignity, and good humor, social, moral, aesthetic, and economic changes more extreme even than any it has known. This is reassurance of a kind we need. How many new houses are efficient, in this way, as tools for living? We shouldn't build another if we owned the United States Steel Corporation.

Outdoors, however, our purse has been more heavily taxed, and our Latin tendencies have had freer play.

When I came out here, a small corn crop was imminent, and the ancient corn crib was on the point of collapse. I located

Mr. Sligo and an old cement man, and they were helped by Hubert Brooks, our hired man at that time. I paid the two outsiders fifty cents an hour, and bought, myself, the rough lumber, cement, gravel, hardware, and metal roofing. The whole thing, which would hold up to five hundred bushels, and is now used for storage of other things besides corn, cost seventy-five dollars. I was moderately impressed, and proud. It worked, and had a certain neatness and honesty. A friend took some modernistic photographs of it—the slats make a nice pattern—and I called it my barnyard Parthenon.

The next thing was the reconstruction of the waterworks. I shan't go into the technical details, but I may say that I had to consult geologists, water witches, engineers, and the County Agent. Finally I decided that wells weren't feasible, and that the cisterns had to be rebuilt, by specialists. As I think I have said, this cost three hundred dollars, and from a hygienic point of view, has not been wholly successful. However, this problem isn't beyond solution, and in the worst droughts we have had plenty of stored water, of one quality or another.

The next problem was an ancient outbuilding, made of rotting wood, that had neither foundation nor windows. It was used for wood, coal, tools, junk, hens to be broken of setting, and anything else that needed some sort of shelter. We tried also to use it for dogs. My Scots blood made me endure it for more than six months, but when I saw how wasteful it was, and when, as an eyesore, it became quite painful, I began to give way. Its doom was spelled by the death of Peggy's first dog. (Haven't we let capitalism do quite a lot of killing, in one way or another?) The year before, Peggy's car had been wrecked; still on salary, she had bought a new one, and then had collected two hundred dollars on the wreck. In a moment of abandon, she had given me this money, and we decided to spend it, prosaically enough, on a new outbuilding.

I made various sketches, and then drew a plan to scale. It included a single garage, a dog kennel, a tool room, and a coal bin. We decided to dispose of worked wood and setting hens elsewhere. Neither of us had seen a real dog kennel, but we designed it as we thought it would be useful and it later appeared that we had hit on the most approved way of building dog kennels. We planned a concrete floor, and kept in mind lighting, ventilation, cleanliness, all the old materials we could use, proportions, the balance of windows and doors, and all the rest. We had never before designed any kind of building, and we found it exciting.

I called Mr. Sligo, and he came out with his partner. We figured the expense of materials as closely as we could, and Mr. Sligo set a lump sum for labor. I usually prefer this arrangement, because it gives me exact figures on cost of materials, and insures speed. Mr. Sligo and I both kept track of time, and at the end found that I had got the job only $1.50 cheaper than if I had paid by the hour. I offered to pay or at least to split this, but Mr. Sligo wouldn't take it. The first estimate was $250, but we managed to whittle this down, even while making alterations in the original plan. The final cost was $209.50, and I have rarely got so much use, value, economy, and comfort for the money. We call this building the Dower House.

The only serious mistake we made was in not slanting the concrete floor to allow drainage in washing the rooms out.

The two yards that Hubert and I built, with great difficulty, for chickens and dogs, back of the Dower House and the hen house, cost about thirty dollars. We used the best close-meshed Elwood fencing, five feet high, but we worked in very wet weather, and didn't get the yards absolutely tight. They have been extremely useful for the dogs, and less so for the chickens.

Last summer, Mr. Sligo and his partner put a new metal roof on the old barn, repaired the floor of the haymow, and lined the part I use for lambing with beaver board. The winter before, the roof leaked on expensive hay, James and I nearly broke our legs by falling through the floor of the mow, and lambing, with icy winds going right through the building like knives, was something of a feat, and involved any old bedclothes Peggy could spare. I didn't like to spend the money on an old building that looks as though it were going to collapse at any moment, but Mr. Sligo assured me it was much stouter than it looks, and that if we ever decided to build a new barn, we could use the new roof off this one. I was glad, not only because I didn't have the money for a new barn, but also because the outside boards of this one look like old pewter.

No farmer should hire this much building and repairing done, but I defy anyone who has anything else on his mind and hands to learn carpentry and concrete work in a couple of years. James and I did close up the end of an old cowshed, put in feed racks for sheep, and make a dozen lambing pens on hinges. I bought the lumber and got the ideas from government pamphlets, and we did the work together. Once in a feed store I saw a brooder made of cardboard and hardware cloth, and warmed with electric lights. It was fairly expensive, so I came home and made one with wood instead of cardboard. Mine worked quite well, and it was larger, stronger, and cheaper, but it was too heavy, and too hard to clean. Also, since we have no plugs in any outbuilding, I had to keep my brooder in the kitchen. Peggy was a good sport about it, but indeed, who was she to complain about a little detail like that? The next summer, when it was 105 in the kennel, she brought Jeanne and the litter of pups right into the library. Short of going to bed with a pig, we are ready for anything.

We dream of a bathroom and a furnace; of tiles in all the fields; of a hen house with incubator, brooders, batteries, electric lights, and a phonograph; and of a new barn with concrete floor, running water, and electric lights. But we don't let any of these or other dreams run away with us, because we don't make capital expenditures until we have the money, and can see it coming back.

Also, because everything in the world seems too uncertain for heavy building and extensive planning. Agriculture and country life may seem as stable as anything in the world, but even if that's true, it isn't saying much. If these extremes of heat and cold, and of droughts and floods, continue, agriculture in this region will have to be profoundly changed, and how, no one knows. Another incalculable element is the progress of chemistry and industry. Rayon is already competing with wool, and vegetables are being grown in tanks of chemicals, on a commercial basis. If climate and soil can be supplanted by less uncertain and inefficient media, they will be, despite social consequences not pleasant to contemplate. It may be that all food will soon be grown in factories, while farms are adapted to the production of raw materials for industry, or simply reforested. Finally, the economic and political situation seems more full of danger, promise, and deep change than at any time since the destruction of feudalism. We farmers may be collectivized, we may be completely isolated and plundered, and we may simply go further into disorganization, inefficiency, tenancy, and dependence, being still more fantastically pampered and robbed.

All this may seem far from building outhouses, but if you will look about you and sniff the air, you will see that it is not. Money—even $209.50—is still a means to freedom, if you have that much and know how to use it. But who does? All decisions, and all buildings, from our Dower House to the Empire State Building, look pretty foolish, if you ask me. All but one: the decision to keep alive, if we can, in more ways than one, and if we have to, to do that, to build a new America.

V. Business

How not to buy sheep.

Most unfortunately, from my point of view, the immigrant farmer can't simply lay his plans, do his work—that is, somehow, by the grace of God, keep his animals alive, and his farm going —and keep his place in working condition. He has to spend a lot of time and energy scheming and figuring, buying, trading, selling, investing, and keeping accounts. Now all this has a certain fascination, as a game, and like that significantly popular and ephemeral parlor game called "Monopoly," it is in the long run a crushing bore. How people spend their lives doing these things is something I can't imagine without great effort. They seem so far from making and using, from life and death, from reality. Well, since I have this ingrained prejudice, I probably can't make these activities, these parts of the life, at all interesting in words. I shall, therefore, try to skip them in their essentials, try to avoid making speeches, and only tell a few anecdotes, and suggest a few characters.

When I first decided to buy some sheep, I consulted my neighbor, Mr. Ralph Stone, and he told me about an old man named Silas Finchley, who lived out in the Huntington hills, and who had been for some years one of the two biggest sheep men in the county. Incidentally, Mr. Stone told me that Mr. Finchley was the best hand at training dogs he had known, and at one time had had a collie that handled sheep perfectly, at the slightest breath of command.

Well, one fine day in August, Peggy and I drove out in search of Mr. Finchley. Registered ewes are generally sold, bred, in November, but the best time to buy ordinary ones is in August, when sheep men are sorting them out and getting ready to breed. First we found Mr. Finchley's brother. He had a couple of lovely colts, but no sheep. Then we went to another farm, by mistake, and found that they had some Muscovy ducks for sale, but no sheep.

Finally, at a gracious old white house in a little valley, we found our man. His brother had told us that he had just arrived home, that very afternoon, from a vacation in California! There he was, in his barnyard, sorting out ewes, and training a collie pup. He still had his town clothes on, and his face, beneath his white hair, was purple with rage. The man he had left in charge had been careless. He was obviously in no mood to do business, and when he told me to come back in a couple of days, we quickly departed.

The very next day, my telephone rang, and it was a woman speaking for Mr. Finchley. He was ill and in bed, but he wanted me to come out immediately and look at a dozen ewes he had selected for sale. I said I'd rather wait until Mr. Finchley was up and about, but no, the woman said, he was very eager to have me come right away. I decided to go, but I felt that I needed a little expert advice, so I called up another neighbor, Mr. Gabriel Oak, who as usual had kindly offered to help me. Luckily he was free that afternoon, and we drove out together.

We were greeted by Mr. Finchley's nurse. She was a middle-aged woman in a nurse's uniform, but she did not look or talk like a registered nurse. She insisted on showing me into Mr.

Finchley's room. The patient was lying in bed in a work shirt. His eyes were sunken, his lips were purple, and he had a most alarming pallor. He seemed to have lost control of his voice, and from his mumbling I could hardly make out that he was telling me how good the twelve ewes were, and why. I told him I'd look at them at once, and left the room.

The nurse showed Mr. Oak and me out to the barn where these ewes were penned, and on the way she explained that Mr. Finchley had been seized with indigestion in the night. She had made him drink blackberry juice, and had put a mustard plaster on his stomach. Evidently this nurse was a neighbor, and no physician had been called in. We made no comment, and went on to look at the sheep. Mr. Oak soon showed me that they were not well bred, that they were very thin, and that their teeth were worn down to the gums. (Any sheep book or pamphlet has pictures showing how to tell the age of a sheep from the teeth.) It was evident that they might be of use to a glue factory or fertilizer works, but not to me.

We went back to the car, and the nurse appeared and told me that Mr. Finchley insisted on my coming back to him before I left. Apparently he had detected my ignorance, had not seen Mr. Oak, and was determined to make one last sale on his deathbed. That it was his deathbed, I at least was convinced, and I shouldn't have gone back into that room for anything. I had visions of his dropping dead from anger and disappointment, and of a coroner's inquest conducted by Dr. Ames and involving this "nurse" and myself. Mr. Oak had disappeared into the car, because he didn't want it known that he had ruined the sale. We got away rather quickly.

I was glad to learn later that Mr. Finchley had survived. Sheep are not likely to provide vacations in California, but if you have a dozen of your oldest and poorest ewes to sell, at a high price, you may be able to rise from your deathbed in hopes of doing so.

Trading and a trader.

Mr. Oak and Mr. Stone are very busy men, and I take advantage of their kindness quite often in the ordinary course of events, so I decided that in this matter of buying sheep, which might take a long time, I'd not trouble them further. Through a friend in town I got in touch with an old man named Job Withcott, who had been a drover in his youth, and was supposed to know more about livestock values than almost anyone in the county. Day after day, Mr. Withcott and I toured the county in my car, looking up flocks of sheep that he had heard were for sale, until we finally got the sheep that both he and I thought were what I wanted. I bought them from a man named Harrison, who had to sell them because neuritis made it impossible for him to care for them. Mr. Harrison had fine, blue eyes that were drawn with pain, he liked to digress, and he moved very slowly and quietly. He knew his sheep intimately, was obviously genuine in his regret at selling them, and gave me much good advice about them. The man from whom I bought my purebred ram, a year later, had many of the same qualities. And from other shepherds I have seen I should hazard the generalization that the breeding and care of good sheep makes good men, or vice versa. If I have sheep for forty years, my mind may wander slowly all over the place, like a sheep, but I may be worth listening to.

But the most interesting character, to me, in all these little transactions, was Mr. Withcott himself. Now Mr. Withcott is a typical trader, if there ever was one, and he has increased my respect for the whole tribe, chiefly by making clear to me the distinction between the typical trader and the typical large gambler or the typical entrepreneur. Mr. Withcott and his kind differ from large gamblers in that they know at first-hand, and like, the things in which they trade. It doesn't make any difference whether these are the livestock, wool, hides, skins, and old tools that are Mr. Withcott's media, or the old prints, paintings, and pieces of furniture that occupy dealers and collectors in New York. To traders these things are not only counters in the only game, but also objects of great interest, in themselves. It seems to me that there is an important difference here between these men and those who gamble in grain futures without knowing a good ear of corn from a bad, or in stocks and bonds without being at all capable of operating a factory. Mr. Withcott and his fellows, whom I encounter at stock sales and elsewhere, usually have small farms of their own, and your good art dealer and collector usually impoverishes himself by holding on to things he loves regardless of their value. Furthermore, your good trader lacks the rapacity, ambition, and religious awe for property of your entrepreneur. He takes too many beatings, and has too much fun, for any of that. He is always much more an actor and a connoisseur than an owner.

Take Mr. Withcott: a large, slow-moving old man, in poor health, with the eye of a hawk and a tongue that never stops. He seems to live alone in town, next to an old barn in which he stores, temporarily, whatever he happens to have on hand. He has been trading for forty years, and hundreds of thousands of dollars' worth of animals and hides must have moved through that barn, and hundreds of thousands of dollars must have passed through Mr. Withcott's wallet, but very little has stuck, and I have a feeling that Mr. Withcott hasn't a penny to his name, in any bank. One reason is that his sole purpose in life is to prove to himself and to others that he has a keen eye for a bargain. He would deny it vigorously, but I suspect that the actual idea of making money, adding it up, and keeping it, has never entered his head effectively. Another reason for his relative poverty is that he is basically too much interested in people, too human, and too kind. Once he has made his point about his cleverness, and once he has found out everything he can about the people he is trading with, his only interest is in being fair and in doing people favors. This must have betrayed him all his life. He went to endless pains to get me good sheep at a fair price, he must have spent forty-eight hours, in all, doing it, and his charge was two dollars.

I got some good sheep at a fair price, I learned a good deal about livestock values, and I saw some excellent comedy, but what I remember most is the biggest earful of gossip, kindly gossip, I have ever heard. Mr. Withcott had traveled all over this region all his life, and had bought, sold, and traded with most of the farmers, with their fathers, and with their grandfathers. In all this time, apparently, he had never forgotten one thing, not the last penny in any deal, and not the slightest shred of gossip about anyone. It is quite beyond me to reproduce that Niagara of talk, but I may be able to give a hint:

"You see that little white house up there on that hill? It belongs to Jane McCoy. She was Jane Saunders till she married old Fred McCoy, along about nineteen-o-five. Old man Saunders was pretty close with his money, and pretty queer. They say he used to get het up about something, once a week, reglar, and write a letter about it to the *Gazette*, on the back of an old envelope, turned inside out. They used to print a letter of his'n about once every six months, but he kept writin' them, once a week, for nigh onto forty years, until he died, along about nineteen-o-nine. He used to pick up pebbles, they say, and carry them in his coat-pockets, and he used to wear that same old coat of his the year round. Jane used to take the pebbles out, and once she tried to burn the coat, but the old man caught her at it, and gave her a tanning with his razor-strop, and her a grown woman, too. Mind you, I don't know how much of this is true; I'm just tellin you what I've heard.

"Well, along comes Fred McCoy. You heard tell of Fred McCoy? No? He was a good one, he was. A timid looking sort of little feller, but keen, and from what I've heard, he had a lot of lead in his pencil, he did. He come up here from Kentucky with nothin but a load of eight mules. Where he got them mules, I don't know, but they was the prettiest load of mules you ever seen, and I bought them, for $105 apiece. That was a high price, then, but I wanted them mules, and Fred McCoy, he seen I wanted them. I was younger then than I am now. He held out for three days for the five dollars, but meanwhile I got a man from Kingston that wanted them mules, and was willin to pay me $125 a head, so in the end I didn't do so bad, and Fred, he didn't do so bad, either.

"Well, one Sattiday night Fred was spendin his money, or some of it, and he come out of that bar down on east Water Street. What was it they called it? I can't rightly remember. You wouldn't know it, anyway. They sold it out along about nineteen-sixteen, just before the War and the camp, when they could have cleaned up with it. Well, as I was sayin, Fred McCoy, he was comin out of this bar, not drunk, mind you, but happy. And who comes along, in that damned old rig of his, with that old white horse that was blind in the off eye, but old man Saunders, with Jane up beside him . . ." And so on and so on, by the hour.

Mr. Withcott says he has a bad heart, and doesn't figure he can live much longer, but somehow I don't believe that. He has too much in his head, and too little in his purse. Men like Mr. Withcott travel very light, and very far, and are hard to catch.

Pieces of silver.

I don't really enjoy selling anything but words, wool, and animals that I can't feed any longer and don't know very well. I usually make it a point not to pay too much attention to the calves, the steers and the grown wether lambs, so that their departure will mean less.

But there is another kind of selling that can be downright unpleasant. For instance, I had to sell a lot of old books that weren't worth anything, and were taking up room. But they had been in that back room for years. I sold them to a second-hand book dealer in town who will never make money, because he is much more interested in contents and associations than in values. He actually persuaded us to keep several books that had been printed here in Chillicothe. He said he was collecting Chillicothe imprints, and thought there ought to be more than one collection!

Another case of the kind was the forced sale of my aunt's old phaëton. It was just about the last one made here, and it was in good condition. It had a lot of memories, but it was taking up room I needed for animals. I sold it for eight dollars to a man named Red Rotch, who is a real horseman, who once ran a small rodeo, who once loaned me a couple of ponies, who is always surrounded by small boys, and who is a thoroughly impecunious and attractive character. I have not seen the old phaëton in use again, and I hope I never shall.

But the worst sale I have ever made was one last winter. As I have said, we have been losing steadily, from wind, lightning, disease, and old age, the great white oaks in the front pasture that gave this place its name, and that have been its glory. Within my memory, more than half have gone, and soon the place will be fairly bare. Not long before she died, my aunt had to have one oak felled, and the brush burned, because of some disease. But since that time, two others have died, and a third is almost dead, from no cause visible to me.

James and I had been working for a long time on the limbs and upper trunk of that first one, on the ground. Finally, after many hesitations and misadventures, I managed to sell the butt log, fourteen feet long, and three in diameter. But the man who bought it, a timid, wrinkled, toothless, and pleasant old man, refused to buy it unless I sold him also the log of another white oak quite dead but still standing. He also wanted to buy every other good tree on the place, including the two huge black walnuts in the front lawn. He was a nice old man, but a damned ghoul.

A few weeks later, when the road seemed hard enough, three men appeared in an old Ford, with a seven-foot crosscut saw, iron wedges, a sledge hammer, and a file. It doesn't take much machinery to destroy, or—as in this case—to clean up after, a great life. I like most workmen, but these lumbermen, hard as nails, were too much like undertakers. James and I went ahead with our sawing and splitting while the strangers worked on the standing tree. I dreaded that last moment, but I wanted to make sure that they did not fell the tree onto some young ones. Besides, except in the movies, I had never seen a really big tree felled, and I had a certain morbid curiosity.

James and I worked away, and eyed the others. There was some uncertainty about how the tree could best be thrown. The trunk leaned one way, but the heaviest branches were on the side opposite. When the saw got about three-quarters through, it was evident that the tree could be thrown clear of anything else. When they had cut almost through, they moved the wedges and started to saw from the other side. The extra man, James, and I withdrew to a safe distance. When the sawyers paused to strike one of the wedges, the top of the tree, at least a hundred feet above the ground, shivered slightly.

"You through?" yelled one man.

"Yes!" yelled the other.

They both jumped back like cats, and looked up. Nothing happened. That tree was standing on its stump, in dreadful equilibrium. Quickly one of the men seized another wedge and drove it in on the upper side, and then jumped back.

With a slowness that had all majesty, all sorrow, all resignation, the top of the tree began to move. . . . Then faster, faster, with proud, defiant finality, and with a murmur and cracking that swelled quickly to a tremendous, crying choke, it fell.

Before King Charles I lost his head, and long before white faces moved west of the Appalachians with their axes, an acorn had fallen and grown into three centuries of beauty. There it lay, a great pile of black debris, to be hauled away and made into planks for ugly little houses, and sawed and chopped up merely to keep Peggy and me warm, and to clean up the pasture.

James and I picked up our tools and came back to the house in silence. A few minutes later, the old man appeared. They could not take the log away for several days, but he wanted to pay me immediately, and to get the sizes of the few planks I had made him promise to save for me. I told him the sizes, and for the logs he paid me $14.55 and then went away. Long before, Peggy and I had agreed to spend every penny received for dead trees on young ones to plant. But I still heard that final scream of the tree, and felt on my entrails the silence afterwards; and I didn't want to go outdoors and see the new patch of sky. And all I had was five pieces of dirty paper, and some pieces of nickel and silver to jingle in my pants.

Come and get it!

But not all of our selling has this mortuary character. Quite the contrary: most of the things we sell were produced for use, and are goods beyond question. When we sell them, we feel that we have earned whatever we may get for them. If we are cursed by the falsity of the market, and by the spread between producer and consumer, one very real compensation, which I appreciate, is this certainty that my work is worth doing. When I sell, I don't have to "create a demand" by appeals to fear and vanity, and by an elaborate structure of lies, and my personality doesn't enter into it. At the low levels, and within the very narrow range allowed, it is a question of sound bone, muscle, fat, and wool, accurately weighed and fairly graded. If you city folks can't buy my food and wool until it has been bought and sold a dozen times, and can't afford it then because you merely work, and don't own the factories, why whose fault is it? Come and get it, and I for one will meet you halfway.

I don't think we need to fear the men who run the factories. The smart ones are just as eager to get on with the job, and get rid of the buyers-and-sellers, and the stockholders, as we are. The only people who really stand between you and me are the saps and suckers who think they are going to marry the boss's daughter or son and make a million dollars. If we can knock some sense into their heads, the papers in safe-deposit boxes will turn to ashes, the slaughter houses and woolen mills and everything else will go full tilt, you will have good food like mine, I shall have good plumbing like some of yours, and everyone will have a job. Everyone, that is, except the buyers-and-sellers who don't know how to do anything else, and who then, by God, will have a chance to "sell themselves," if they can.

I speak with some heat, and I mean what I say, because every time I go down to the stockyards with something to sell, I see better men than I am who have nothing but their livestock to sell, and who haven't a chance in the world to work up the "specialty" that I know to my sorrow is absolutely necessary, right now, for decent life in the country. They haven't any typewriters, and they haven't any papers in safe-deposit boxes. And I'm not just being altruistic and sentimental, either; I am too close to their situation, myself, for any of that. We want to earn a decent living for goods produced, and we want to get our meat and wool to you, and to you alone, for whatever it is worth to you.

When are you going to come and get it?

A real stock exchange.

Now that, despite my best intentions, I have made this little speech, I can go on to the actual processes and scenes of selling farm goods.

Of our garden produce, we eat about a third fresh, can about a third, and give away the rest to James and to our friends. I have sold a few odd lots, but it is impossible to sell vegetables profitably unless one raises them in large quantities, and ships them, or raises vegetables in greenhouses out of season. Every spring until this one we have shipped to different members of our families forty or fifty dozen eggs, but we can't sell eggs profitably until, if ever, we decide to produce them more scientifically. Now and then I have sold a crate or two of old hens and more or less young cocks to a poultry house in town, but without larger scale production, and without a contract for dressed fowl with one of the meat shops, the profits here are not impressive. The poultry houses pay cash, but aren't very critical, and we shan't ever be interested in canned chicken. Dressed ducks are another matter. We have shipped a few in dry ice to friends, and we may some day find it profitable to sell them dressed in autumn and winter. James is a good hand at dressing fowl, and I can just do it.

As I have said, I sell my wool through the pool of the Ohio Wool Growers Coöperative Association, for which the Farm Bureau acts as agent. I had 114 lbs., last year, from 17 sheep, averaging 6.7 lbs. a fleece, sold the lot on April 30th, receiving an advance of 23¢ a pound, which, less 29¢ commission to the Farm Bureau, made $26.45. A day or two later, Mr. Withcott called me up and offered me a flat 32¢ a pound, one cent over the market at that time. I should have liked to sell to Mr. Withcott, because of his cheap and good services to me, but I wanted my wool graded, and I am something of a fanatic on coöperative buying and selling. This wool was sold on December 3rd, in four grades, mostly ⅜ blood staple at 37¢, with a commission of 2¾ ¢ per lb. My total receipts were $39.82. This year I have pooled 229½ lbs. from 26 sheep, averaging 8.8 lbs. a fleece, receiving an advance of 33¢ a pound, which, less 57¢ commission, made $75.16; the total receipts may be $100.

The grading of wools is a complicated and interesting matter about which I do not yet know a great deal. It is enough to say here that I hope, with better breeding, to average up to 10 pounds a fleece, and to raise the percentage of ⅜ blood staple to 75 or more, instead of 57.

There are two footnotes to this transaction: (1) Last May, Peggy and I had a little twenty-four-hour bat in Cincinnati, which cost us just about the advance on the wool clip; and (2) Peggy's brother-in-law sold me 3½ yards of fine worsted, for a suit, at $2.50 a yard, when he could get $3.00 for the same; and my tailor told me he would have to pay $6.50 a yard for that material. This worsted manufacturer is trying to grow vertically in both directions. Evidently in some kinds of manufacturing, too, the only hope is to reach back towards sources, and forward towards use, and to buy and sell as little as possible.

The reader may excuse me if I repeat here that in our selling of grain from the farm at Bellbridge, special market advices, available to practically no farmers, have been equivalent, in money, to the enormous labor, skill, and knowledge necessary to raise a couple of hundred additional bushels of corn or wheat. Now and then I have been careless about these reports, to my own and my sister's loss. They are written in a very peculiar argot, and their importance irritates me.

In the last three years, we have sold thirteen head of cattle, twenty wether lambs, and one ram. The average price for more or less finished steers has been $7.95 a hundred, and for cows and younger cattle off grass, $4.40. The number sold is too small to make the averages significant, but they may give the immigrant farmer some idea. Lambs are judged more closely than cattle, and it is important to sell them as early as possible, and as soon as they reach the right weight. Here they like them best at 80 lbs. a head. I have no scales, and have to guess. We used to pick up Peggy's niece, and then quickly pick up a lamb. Last year I sold my 12 head on July 3rd, as follows; 9 head, 710 lbs., @ $ 10.25; 2 head, 110 lbs., @ $6.00; 1 head, 65 lbs., @ $8.05. My net receipts were $81.18. I knew that three were small, but didn't want to sell so few in two lots. I figured that I lost about $3.00 by waiting that final week, although the drought kept the price of lamb up, while lowering the price of ewes. I shan't soon forget that in the drought of 1930 you could buy any number of good ewes out here at two bits a head. I sold my first ram at the yards for $8.25, less 29¢ charges, and probably could have sold him the next day privately for $10.•

Our most exciting sale was our first, more than two years ago. We had been feeding a big white steer that was about two and a half years old, and that weighed, I guessed, about 1,100 lbs. I was not yet convinced that I could do better at the auctions, so I got our own butcher, Joe Bell, to come out and look at him. He asked me the price, and I asked the top at the time, $8.50 a hundred. He said that was too much, and advised me to pen the steer more closely, and feed him more. I did so, for another month, and Bell came out again. He still wouldn't buy, and wanted me to feed another fortnight, but corn was 85¢ a

*Later bulletin: of my twelve wether lambs this year, I sold eight on July 9th, as follows: 4 head, 280 lbs., @ $10.50; 4 head, 315 lbs., $11.15; net receipts, $62.18. These were the top prices for the day.

bushel, and I decided to sell the next Friday at the auction. Bell told me I'd get about $7.50 a hundred.

Naturally, Peggy and I went to the sale. It was our first, on a day so cold that it jelled the fluid in one's spinal cord. The scene fascinated us from the start, and we must have visited that place, with and without cause, thirty or forty times since then. The auction we use most is at a stockyard next to the railroad yards, every Friday, and you can tell that day of the week, in town, by the number of trucks with bawling cows and calves, bleating sheep and lambs, screaming hogs and pigs, and sometimes even odd lots of horses, goats, and dogs.

The men are all hard but friendly specimens who mean business, and don't even bother to look twice at the few women who appear every week. (At one time, a rather chic woman was down there buying a couple of thousand dollars' worth of cattle a week.) We always see Dr. Ames, once Mr. Withcott, sometimes our neighbors, and nearly always one or two other men we know, odd characters who interest us more than some of the people we are supposed to call on. There is always one man there who always wants to buy one of my "big red cows," for nothing. Once a sinister-looking youth asked me whether I could locate for him, on commission, herds of Hereford steers running about 1,200 lbs. a head. This surprised me a lot, not only because I didn't fancy I looked capable of buying cattle on commission, but also because "baby beef," of eight or nine hundred pounds, are the vogue. Often Red Rotch appears and wants to "lend" me a couple of horses. Most of the men are shy, and we are too, but we generally pick up an interesting conversation or two. Besides, we know of no better way of learning livestock values, and we always bet with each other on our guesses of weights and prices.

The auction takes place in an amphitheater seating several hundred people. This was a ring, in the old days, before a fire wiped the place out, the night before a rival auction and yard opened. The live-stock are driven in one door and out another. The auctioneer, a large, genial man, shouts at incredible, unintelligible speed into amplifiers, from one o'clock in the afternoon to between five and nine in the evening. He has several stooges who handle the stock and pick up the bids. There is a good deal of horsing and wisecracking that I am beginning finally to understand, but the bidding itself is almost as reserved as at the Anderson Galleries in New York. A flick of the eyelash will buy a big red bull, and I always sit there afraid that the conversation and betting between Peggy and me will result in our having to explain that we really didn't want that herd of mangy heifers, or something. We used to be protected by being total strangers, but some of those alert stooges know us, nowadays. One element in our pleasure, of course, is the excitement of wondering whether we might not do well to buy that steer, going so cheap. Peggy swears that some day she is going to come home with a load of calves, and pay for all feed and labor herself. In short, the stock auctions are one of our major social pleasures, and we drag most of our guests to the yards, whether they are interested, and can stand the smells, or not. People seem to be interested in the New York Stock Exchange, and I may remark, without launching into a speech, that ours is a good deal better.

Well, that first cold afternoon in March, Peggy and I sat there entranced for five hours before our steer came in. Joe Bell sat directly opposite us. Finally, in came that big white steer, mad as hops, and filthy, but looking good. They started him at $8.50, and almost before we knew it, he had been run up to $9.30 and sold. Joe Bell's jaw dropped, and our hearts leapt in our chests. It was as good as any winning touchdown I have ever seen. We got some coffee into our bellies, said good-bye to our steer, and hurried home rejoicing. The next day I "hap-

pened" to have to buy some meat at Joe Bell's shop. He happened to be there, and I tried to act casual about it.

Three anecdotes for ideologists and reactionaries.

The first is from the experience of a good friend of ours in
town, an engineer named Ed Brown. Now this Ed Brown is one
of the few people I can turn to, in my thoughts, when I begin
to fancy that everyone in this country is a crook or a boob, and
that the world is going straight to hell. He is a good, realistic
socialist, in feelings, as well as ideas, he is interested in everything, he uses his head, and he is a first-rate wag and mimic.
His wife is as good as he is, and they are raising a family with
hope and pleasure. He used to work for a gas and electric company controlled, ultimately, by J. P. Morgan. Well, as Ed says,
he got rather tired of telling lies to nice people for Mr. Morgan,
so he got himself a job with less money and more fun, with the
Resettlement Service.

But the story is about the other job, where, too, apparently, Ed was always trying to convert the men working with
and under him. One day he and some of his men were working
on a gas line flooded by a sewer. Ed was on the street, and one
of his men was working in the hole, up to his waist in sewage.
They were talking politics, as usual. The man in the hole had
the last word, for the time being.

"Hell, Ed," he said, "Morgan gimme this job, didne?"

The second story is from my own experience, and less vivid,
unless remembered in relation to other facts, fancies, and opinions in this book.

Last fall we had run out of corn here at Oak Hill, and had
to get some, earlier than usual, from the farm at Bellbridge,
where they had not yet begun to husk. I called up Mr. Kincaid,
and arranged to drive up the next day, with James, and husk out

about eight bushels ourselves. When we got there, Mr. Kincaid was working at something else, but came and helped James and me. There was only about a bushel to a shock, and there was a high wind, and as Mr. Kincaid observed sardonically, it took the three of us a couple of hours, almost, to get out only eight bushels. It comforted me a little to notice that I was only about a third slower than the two experts. I don't imagine that Mr. Kincaid noticed a little fact that meant something to me: for the first time in more than a century, for the first time in five generations, one of the owners of that farm was using his hands on it.

Well, we all talked together, off and on, about the weather, prices, prospects, feeding, and so on. Now I rarely talk politics with either Mr. Kincaid or with James, because they know roughly where I stand, they are shy of politics, with me at least, and I don't like to rub things in, either on myself or on them. They are both Republicans. Well, that day Mr. Kincaid was saying he was having a hard time getting help for husking, because all the loose men were afraid to give up their jobs on the W.P.A. (Husking is piece work, and seasonal, and I understand that the W.P.A. has tried to adjust to it.) I didn't say anything, but I caught Mr. Kincaid and James smiling at each other.

"One of these days," said James, "I reckon these Democrats will be telling us everything we can do and can't do."

"Yes," said Mr. Kincaid, "I figure we might as well have a dictatorship, or whatever they call it, and be done with it."

"Yes," I said, "I reckon you're both right, at that."

But we all laughed together, and in that kind of laughter, I think, lies the hope of my native land.

That's why I don't mind countless little episodes like these, outdoors, at work, with men whom I respect and to some extent understand, and who respect and to some extent understand me. But there are other little episodes, indoors, with other men and women, when the job on hand is the shuffling of pieces of paper (with knaves, kings and queens, and figures, or gilt edges and allegorical and arabic figures), and when the mutual respect and understanding are considerably less. The scene of my last little story is indoors, in a bank. Before I tell it, I should like to say that the officers and employees of this bank have been notably courteous and helpful to us both.

Now I don't suppose that I have to repeat that my experience as a picayune investor will have been brief, involuntary, and extremely lucky, and that I am extremely grateful for the expert advice on which I have acted completely. Neither do I have to explain why I think of this part of my private economy as a highly inflamed appendix of which I hope to be relieved painlessly. I may, however, be permitted to add that I find the actual process disagreeable, simply because it warps and embitters my view of human nature. Every time I go into that bank, and get that little steel box, and sneak into a cubbyhole with it and a pair of scissors, my mind is full of images of men and women and children, in steel mills, lead mills, power plants, packing houses, foundries, railroads, telegraph offices, and so on, all over this country, and even abroad, sweating their guts out, asking for raises in pay, and not getting them, simply because a lot of people are doing what I'm doing, and have to be paid first—for *what*?

It is all very unpleasant, much, much less pleasant than shoveling dung or packing corn fodder in an icy rain. I used now and then to try to relieve my bitterness by joking with the clerks in the bank—sympathetically and sardonically taunting them, as a matter of fact. But my jokes are apt to be a bit too tentative, elliptical, and obscure, and one final effort in this direction brought an end to them all.

I came out of my rat hole one day with a handful of coupons worth a couple of hundred dollars, and gave them, with a deposit slip, to a teller, who was probably earning about thirty dollars a week, and who knows how I got my tiny wad. I was in my working clothes, and my hands, though not like James's or Mr. Kincaid's, do not look exactly feeble. The teller gave me the usual sheaf of papers on ownership to sign. Taking them, I assumed a grotesquely exaggerated attitude of weakness, bore- dom, and annoyance, and said: "What a beastly bore! Why do I have to sign all these things? Why do I have to tire out my hands and fingers using a pair of scissors?"

"I know," the teller said, sympathetically and seriously, "I know! Just one bit more of government regulation! We're all tied hand and foot, these days!"

I muttered a few odd blasphemies to myself, and came away.

The red and the black.

Bookkeeping and accounting do not rank, for me, among the pleasures of farm life. Peggy and I keep track of income and expenses as well as we can, and then every month beg and taunt each other into a joint accounting that takes several hours. Once a year we have a bout of accounting that lasts days, and then shadows our lives for weeks and months afterwards.

Often enough I ask myself why we do this. Obviously, I have to keep track of income for tax purposes, whether to pay heavily and gladly for ill-gotten gains, or merely to prove that our exemption as married people lets us out of income taxes en- tirely. Also, I am, you might say, in business, and have to know, if possible, where I am going, and why. Finally, my sister's part ownership of this place and of the other farm lets her in on up- keep expenses here, and on gains, if any, there. In all these re- spects I am, with herculean efforts, fairly accurate, and see no escape.

It is our personal and household accounts that really hurt, and that raise the question. Our income and expenditures are so irregular and uncertain that a budget is quite out of question. We are two of the few people who have not fallen into a clinch in the moonlight and protested: "But we can make a budget!" Still, we kept fairly accurate accounts for 1935, and these helped us, a little, to cut our expenditures for 1936. Our accounts for 1936 may do the same for 1937. Yet I doubt very much whether the savings were worth the mental wear and tear of complete accounting. Right now, Peggy is urging me to get to work on the books with her, but I have a horrid suspicion that our personal and household accounts will be extremely rough from now on. We have slaved at them enough to have some idea where the money goes, and here the law of diminishing returns seems to be active.

The only good result of the whole effort, that I can see—and this good is rather doubtful—is that I am now in a position to make with assurance my warning that small farmers, and especially small farmers from the city, simply must have a "cash crop," a trained and marketable ability, a specialty. Long thought, with pencil in hand, must precede any acceptance of my invitation to the country.

Now for a little evidence to back this warning, a few odd hints and pointers, from our own experience. I have here a ten-page Appendix, full of figures, that I have prepared, with great effort, for this book, and decided to omit, for three reasons: (1) The typographical composition would be so expensive as to dishearten my publisher; (2) it would be read only by gossips who know us and would pass it on, distorted, to others; (3) I trust the instinct that makes it easier to sleep with a woman, or to learn a man's most intimate ideas, than to learn the essential facts and figures of his or her private economy.

As I have said, we'd like to have an income of about three thousand dollars a year. In 1935, despite wedding gifts and a fairly good year at Bellbridge, our "income" fell, you might say, fairly far short of that figure, and our expenditures ran about twice as far beyond it. Peggy spent less than her small income. The running loss on Oak Hill, chiefly wages, was about six hundred dollars; we had to spend more than five hundred on permanent improvements, like the cisterns and electricity, and we had to spend nearly that much on household repairs and gadgets. Groceries and cars ran about three hundred each. Domestic wages (including laundry), insurance, and health, ran to about two hundred and fifty each. Amusements and myself came to about a hundred and fifty each. Coal, liquor, books, and the egregious "miscellaneous" came to about a hundred each. All the rest, electricity, dues, Christmas and other gifts, garden, express and P.O., three dogs, meat, tobacco, personal taxes, and telephone and telegraph, came to much less than a hundred each. At Oak Hill, excluding labor, I made about forty dollars on the cattle, and we got about one hundred and sixty dollars' worth of what we call "credits" on eggs, chickens, butter, milk, and cream. The garden saved us about fifty dollars. Our capital was reduced about 18 per cent.

In 1936, thanks almost wholly to the fantastically lucky transactions already mentioned, our income rose above the ideal figure mentioned. Thanks to greater experience, and to capital goods acquired, rather than to any painful economies, our expenditure fell much closer towards our income, though stopping uncomfortably above it. The net loss on Oak Hill fell slightly. The drought raised our grocery bill. On account of guests and ill health, domestic wages rose. Thanks to Peggy, and despite a new electric stove, household expenses fell considerably. Cars, health, and insurance about the same. Taxes, dogs, and amusements rose. Both Peggy and I came a little cheaper. Electricity rose, but coal fell even more. Books and liquor fell off noticeably, and everything else was about the same. The income from Bellbridge (the debt, rather, to Mr. Kincaid) was less. On Oak

Hill, the upkeep included termite treatment, and rose. The hard winter raised feed bills. Again excluding labor, I made about a hundred and twenty-five dollars on the cattle, and about forty-five on the sheep. Despite the drought we raised our produce "credits" about forty dollars. Our capital reduction was about 4 per cent.

This year we shall probably do much less well with Bell-bridge and investments, and much better at Oak Hill.

It is to be noted that Peggy pays for her own clothes and personal expenses, and that our families, friends, and guests have given us large quantities of food, liquor, clothing, and everything else. We have "discovered" what everyone knows, that most of the departments of expenditure mentioned account, each, for less than 5 per cent of the total expenditure, that is, that the little things are what add up. We think that we can economize further, as we have done, rather painlessly, down to about three thousand dollars a year. Reductions below that figure are also possible, of course, but would mean a change in the quality, as it were, rather than the quantity, of our life, would mean major alterations in the way we pass our time, the social relations we have, and our general attitudes. This prospect doesn't frighten us especially, except in one way. We aren't old, or even middle-aged, yet, but neither are we as young and strong as we were ten years ago, and even in our present luxurious style of living, fatigue is our major enemy. And with the best intelligence we can muster against it, it will not grow less dangerous. Once you are beyond thirty, even simplification demands energy, and to risk a largely irrelevant thought, Thoreau died at forty-five.

VI. Bodies

The incompleat gardeners.

I hope I am not laboring the obvious when I point out here that in the country the processes of keeping well fed, clean, warm, and comfortable form together a major part of one's life. They are absorbing, and both difficult and pleasant. They are closely related to production and business, and to pleasure. Here, one is less dependent on society, and more dependent on nature.

In this general realm, the most obvious necessity and advantage is a good garden, or at least a garden as good as one can make it. Here I have to make a rather damaging confession: I lack that real passion for gardening that would seem to be essential in any true enthusiast for country life. I have worked a good deal in our garden, with pleasure, and I certainly enjoy the vegetables and flowers, but I am given away by the fact that my interest comes and goes. When the garden is almost choked out by water and weeds, or burned up by the sun, or devoured by bugs, I begin to lose interest, and towards the end of winter it is Peggy who begins to read the seed catalogues, and to make the plans. In fact, Peggy is much closer to being a real gardener than I am. She knows exactly what's going on out there, at almost any time, and has that limitless ambition, that light in the eye when the subject comes up. But she is not a real gardener, either, because her interest has its peaks and valleys, too. She may be one, some day, while I have almost given myself up. I take some small comfort in the fact that the best gardeners I have known have not been farmers, but suburbanites or professors, and that most farmers, including the best ones, are no more interested in their gardens than I am in mine.

Still, I regret this, because my grandfather and aunt were real gardeners, and the old garden here really deserves a couple of fanatics. It is not, and never has been, a garden to make people stare and exclaim, but it has its own charm and possibilities. It is about half an acre large, and is divided by stone walks, put down when the house was built, into six rectangular beds. There are trees near it, outside the fences, and some young ones inside. Until it was blown down, an old apple tree at the far end of the center walk established a focal point now lacking. On either side of the walk leading to the garden, there are five large old lilac bushes. My aunt had all her flowers round the edges of the beds, and so, although she raised some fine ones, they were usually concealed by the vegetables. Peggy has cut about twenty feet off the end of each bed, where it abuts the center walk, and is in the process of moving most of the flowers to these central beds, which together make a vista. It will take many years to get everything moved, and to get all the bushes and perennials in order, as a background to the annuals, and as a green wall separating flowers from vegetables, but when she is through, the effect will be good.

For many years, the garden has got all the manure from the hen house, and some from the barnyard, with bone meal and other delicacies for pets, so it is rich. However, like all the soil on the place, it is badly drained, heavy, apt to be sour. We have used a good deal of lime, and intend to use more.

The first summer I lived here, Hubert Brooks did all the

gardening, superficially as it had been done, but not nearly so well, because my aunt had always stood over him, and told him precisely what to do. He had a fairly good season, and some good vegetables, but the flowers had suffered from a hard winter, and suffered still more from the ignorance and indifference of both Hubert and myself.

The next summer, Peggy was here, we both got interested in the garden, and some high comedy ensued. On the whole, Peggy knew more about gardening than either Hubert or myself, but there were some essentials that she didn't know at all, and this destroyed both his confidence in her, as a gardener, and mine. Hubert claimed to know everything, and we made the mistake of asking his opinion on this and that, and acting on it, before we discovered that he knew practically nothing. I knew nothing, and knew that I knew nothing, but like Peggy, I felt I had a right to make my own mistakes. Peggy grew bossy, Hubert grew surly, and I grew disgusted, while the weeds merely grew apace. Hubert and I both had a lot of work to do outside the garden, and Peggy spent August on Cape Cod. It was a very wet summer, and in the end, it was the weeds that had their way.

By the next summer, Hubert had been replaced by James, and we got a fresh start. James knows a lot about vegetables, and nothing about flowers. He knows what he doesn't know, does cheerfully what he is told, and gives advice only when it is asked. Our neighbor Ed Wagner plowed the vegetable beds in April, and did an excellent job. The soil was in good condition, and we got away to a flying start. All three of us worked hard, and we got our spading done, and our first seeds in, early. They came up well, and we kept the beds immaculate, if I do say so. We had a lot of fun, and by June that garden was our pride and joy.

A great many things, including all the rambler roses and grapevines, had been winterkilled to the ground, but many had started to grow again. We took this opportunity to take down the old rose arbors and build new ones. My aunt was a very small person, and unwittingly had them built so low that only she was able to walk under them.

This, we thought, is gardening, and maybe we are gardeners after all! . . . Then it stopped raining completely, and for weeks on end the temperature ran above eighty, ninety, and a hundred. The plants stopped growing, and withered; the vegetables were baked, premature, and tasteless; the flowers were stunted and colorless. For weeks, I carried sixty or eighty buckets of water every evening or so, but this effort proved quite vain. Slowly and surely, almost the entire garden was wiped out.

Fortunately, our minds were kept off it: there were the pasture, meadows, and livestock to worry about, too, we had many guests, and finally (miserable distraction!) Peggy spent six weeks in bed, gasping with asthma. But in the autumn, after all the guests had departed, and the asthma had begun to be cured, we got a few showers, and began to take an interest once more in the garden. Most of the vegetables were gone, but the autumn flowers had brilliant color, and it became possible to do some transplanting, to clean up, to spread the manure, and to look forward to another year.

This spring has been warm and damp, and everything has burgeoned wonderfully, but we have not forgotten that drought.

Salvage.

In the spring, last year, we had quantities of lettuce and radishes, my grandfather's old asparagus bed, with some replacements, did nobly, we had strawberries daily for several weeks, and there were moderately good crops of peas, green beans, carrots, and beets. If there are any things better to eat than asparagus, strawberries, and peas, eaten within an hour or so of being cut or picked, I don't know what they are. We didn't plant the strawberries right, and they have run a bit wild, but we are gradually getting them under control. I tried both bushing the peas and running them on strings between stakes, and favor the latter method. We planted one bushel of Irish potatoes, and though they started well, the drought got them, and we dug only five bushels, one of which went home with James.

The Golden Bantam sweet corn tasseled out between two and three feet high, and we didn't get more than five dozen poor nubbins, but that was better luck than some of our neighbors had. The cantaloupes started famously, and then, although I carried what felt like tons of water to them, were completely ruined. (Some years excellent cantaloupes have been brought in from that garden by the wheelbarrow load.) The lima beans stood the drought better than anything, and we had an almost steady supply, for weeks. Spinach, shell beans, and onions were poor to fair. We put in some cucumbers, and then, because we feared their crossing with the cantaloupes, dug them up. James quietly saved a few in a remote corner of the garden, but these were burned up too. The tomatoes, more than anything else, broke my gardening morale. They started luxuriously, and I staked, tied, and suckered them most carefully. When the drought came, I carried quantities of water to them, too. When the tomatoes were an inch or two in diameter, they began to rot and bake on the vines. We got a few buckets full of clean, small tomatoes, but their flavor was poor. We had a fairly good supply of broccoli, but nothing to what we should have had. The year before, we had had some inferior Brussels sprouts: I had not thinned them enough, or dusted them properly. We have some old currant and gooseberry bushes, but they have been attacked by things we cannot diagnose or cure, and we have had no berries. Last year we set out some raspberry roots, and it looks as though we'd have a few this year. Every year we put turnips in the beds that have been used. We don't like them, but James does, and they make a little feed for cattle and chickens.

On the whole, we have had relatively little trouble with bugs. (Compared to any garden in this region twenty or thirty years ago, we have been cursed.) We use a good deal of a patent "Rotenone" dust that is variously useful, and some Bordeaux on the potatoes.

We try to can as much food as possible, but so far, we have not done remarkably well. The first year, while Peggy was away, I made Hubert's wife can forty quarts of tomatoes, a bushel of bought peaches, and fair amounts of lima beans, blackberries, and so on. She was a good canner, and the things kept, and cut down our grocery bills, but we have not yet made good the investment in jars and equipment. The next year, with Peggy in bed, a new girl on the job, and little coming in from the garden, the results on the shelves were not impressive. I remember especially a day spent by this girl and me, canning lima beans that were mouldy within a fortnight. When Peggy was well, we dried some limas successfully. Some of Peggy's relations have the good sense and generosity to give us canned goods for Christmas. I never thought I'd look at shelves of canned food with the feelings with which I have always looked at shelves of books, tobacco, and liquor.

But it is as hard to talk about a garden as an economic affair as it is to put a price on a dog. Think of the freshness of the vegetables, and of one's ridiculous pride in them. And, most obviously, think of the flowers.

The society of flowers.

These mean more to both Peggy and me, perhaps, than any or all of the vegetables. I am hopelessly ignorant about them, as I am about music, but they mean quite as much to me as music does. I can't even remember more than half the names, attaching them to the right flowers, so I have had to get a list from Peggy. This will mean something to garden fans, and others can skip it. I started to put them down in their proper order of flowering, and to distinguish between perennials and annuals, and to put them all in the singular, or all in the plural, but I had to give all that up. My list will have to stand as poetry, or as a roster of good companions. Here it is, for whatever it is worth:

snowdrops	hollyhocks
daffodils	herbs
narcissi	amaryllis
squills	cosmos
grape hyacinths	marigolds
crown imperials	nicoteana
iris	scabiosa
lilacs	zinnias
honeysuckle	petunias
tulips	coreopsis
black currants	bleeding hearts
flowering almonds	portulaca
snowballs	sweet alyssum
peonies	candytuft
lemon lilies	nasturtiums
sweet rockets	shirley poppies
roses (20 or 30 kinds)	oriental poppies
pyrethrum	asters
cornflowers	phlox
star of Bethlehem	gladioli
larkspurs	flax
columbines	mignonettes
mallows	morning glories
calendulas	verbenas
foxgloves	

There are daffodils and narcissi in the garden, but there are also thousands of them along the fence of the front yard, and here and there under the trees. I think right now that I like them better than any of the other flowers. They come at a time when flowers, any flowers, seem absolutely necessary, and they are not just any flowers; they have such marvelous freshness, such youthful distinction. Of the iris, I like especially the wild and imported ones, with narrow petals. The peonies are like a tea party of large, handsome, opulent young women; grand fun if you can browbeat them and keep them in their place. A rose is a rose is a rose; sometimes I think they are the ultimate in flowers, and often they seem a bit too much. Of all the summer flowers, I think I like the cornflowers the best. Last year, throughout the drought, almost, we had quantities of them, of the most clear, deep, and perfect blue, and we put them in clear glass bowls. I don't know, though. The scabiosa are more subtle, and zinnias are as real and satisfying as if they had been painted by Van Gogh. And the zinnias always remind me of those last few days with my aunt. . . . But I'm not going on with this gossip about flowers. You don't have to be a real gardener, you don't even have to know their names, to feel them, in a garden or in a room, like people. Often enough, they are better company than most people.

How to be a gourmet, though relatively poor.

Even more important than a garden is a wife who is an intelligent and thrifty hedonist, and who sees a kitchen as a studio. My few notes here are essays in a form of appreciation. In this respect I began as a model husband; I had eaten enough good food to appreciate, but had not learned enough to criticize. Peggy is slowly and surely turning me into a critic, and will have to take the consequences. A half-baked critic is as bad as a half-baked potato.

Despite our ups and downs in canning, there have been few times, in winter, when we have been quite unable to resort to the pantry shelves. The two things about canned food are the freshness with which it goes in, and the ingenuity with which it comes out. I am thinking especially of a simple and superlative winter dish, out of a jar, called escalloped corn. Our most constant grocery expenses have been, perhaps, for canned dog food, for bread, and for fruit. We both like hot breads of all sorts, for supper, and Peggy is now on a no-wheat diet, but we still have to stop in town at night, at an all-night grocery, for bread and oranges. We are both fruit fiends, and though we can raise and steal the fruits I have mentioned, fruit remains a major expense. The prices seem to be higher here than in New York. In the winter we buy tree-ripened oranges and grapefruit, by express, from Florida. The price is about the same, and the quality is much better. Meat is relatively cheap here, and if you get a butcher or clerk who knows you, it is apt to be very good. Neither of us has learned yet to distinguish all the cuts, or to be unfailing in distinguishing a good piece of meat from a bad. Of course our own milk, cream, butter, and eggs (when we have them), make a tremendous difference. We eat a good deal of cheese, but we haven't found the variety and quality we like. Too much of the stuff is badly processed, or expensive. It is very hard to get good fish, and we miss it keenly. We eat tons of salad, and sometimes I feel as though I were turning into a leaf of lettuce.

Peggy's major stores of recipes are the *Boston Cooking School Cook Book*, and a collection of old family recipes made for her by one of her aunts. There are also several old recipe books from my family. In addition, our friend Miss Mary Bell, and other good cooks of this region—where people have always treated their stomachs with respect, at home—have given her some corking recipes. Miss Mary's masterpiece is, I think, chutney. Peggy gets recipes from magazines, especially from *Successful*

Farming, and once our friend Gerald Rowan kept clipping from some magazine a series of recipes from the embassies in Washington.

The general result is that our meals suggest both New England and the upper South, with variations from France and elsewhere. I remember especially one triumph, called Chicken Alexis, invented by a cook at the Rainbow Room in New York. Peggy can produce this more cheaply without debasing it. It is wonderful to me what a touch of wine, even bad wine, or a suggestion of garlic, or of something else one wouldn't expect, in exactly the right place, can do to make an ordinary dish an experience and a memory. It also interests me to notice what playful care, without fussiness, Peggy takes in considering, in relation to each other, the colors of foods and of dishes. When we got the electric oven, Peggy had a few struggles to get used to it, but now that she has mastered it, she swears by it. It is especially good in preserving the juices of meats.

One of the best cooks in this region, on a large scale, at least, is the manager of the farm where our Hereford bull lives, and on our recent visit to New York it shamed me to notice that many of the men we visited could cook as well as their wives, and in some particulars, better. I have always thought I wanted to learn how to cook myself, but I am beginning to doubt it. Several times I have had to do my own cooking, and have not taken advantage of the fact. After a day's work, I have always been willing merely to boil some vegetables or to open a can, fry some bacon and eggs, and make tea, and let it go at that. Or else I have dodged the whole thing by going to town. This is generally a mistake, because the good cooking of this region is done in homes only. The restaurants are either downright bad or fairly good, but lacking in imagination and finesse. I can make the one salad dressing that any restaurant waiter in France can make, and Peggy always has me make it at the table, even when we have company. I enjoyed this little pose until I discovered from the magazines that it was being taken up by old gentlemen in dinner jackets and pince-nez.

How to be an amateur, in the same condition.

The simple fact is that it is not yet possible—and I don't see how we Americans can ever become really civilized until it is possible—for poor and nearly poor people, all over the country, to have a half-bottle of decent wine with every luncheon and dinner. This seems to me quite as necessary for the soul as daily evacuation is for the body. And despite all the discussions of cheap American wines, by the *amateur* of the *New Yorker*, and other experts, we have not yet taken this first step in mental hygiene. The only hope that I can see is in a gradual increase in home wine-making, which would lead to finer discrimination, the growth of a good market, and the establishment of thousands of good small vineyards, each unique, and each proud of the wine that it alone could produce. This may be nothing but a dream. All we can do, meanwhile, is make our own.

There used to be half a dozen good vines in our garden, and for forty years my aunt made, almost every year, a little wine. On account of her heart, she tasted her wine only when she bottled it, and on account of the already numerous temptations of the young, she offered it only to gentleman callers more than seventy years old. When all of these had died, she gave her wine to her church, for communion, and to the hospital, for patients. (How much of it reached the patients, I shouldn't care to know.) Most of what was left, including the best, was consumed one year, surreptitiously, by the husband of a cook of the time.

After my aunt's death, the vines were neglected for a while. They were not pruned properly, the supports were allowed to rot and collapse, and the winters of '33–'36 gradually killed them all to the ground. They have begun to grow again, and we are very slowly getting around to learning how to care for them properly. But before their lowest point, I got enough grapes to make a little wine, according to the old recipe, in '34 and '35. The process is excellent fun, too commonplace to report. It is also very mysterious, and full of surprising accidents, not always bad in result. I promise myself that I shall learn more about it, some day, but I probably never shall, unless the accidents become more unhappy than they have been.

Oak Hill 1934 turned out surprisingly well: a delicate rose in color, very clear, slightly sparkling, nice in flavor, but rather too light and delicate. It was very dry, and not at all acid. I have a hunch that it is moderately stable, or whatever that term is. The 1935 wine seemed, on bottling, to have more character. Its color was deeper, and no less clear, and its flavor stronger, somewhat suggestive of a Beaujolais. There was no sparkle, and it was highly alcoholic. I think I have left four bottles of each year, and I hope to save them for a long time, although I fear that the 1935 is not too reliable. In 1936 we had no grapes of our own, and illness prevented us from buying grapes and making wine.

Spirits seem to me largely a matter of psychology and personal history, as well as private economics. Our experience has been that here on the farm we need a stiff drink much less frequently than elsewhere, but that when we need one, we really need it. On such occasions it is more important than food. When we are too cold deep inside, and too tired, to think of cooking and eating with any enthusiasm, and are getting touchy and quarrelsome, there is only one answer, a good stiff drink. If I had plenty of money, I think that about the only hard liquor I'd drink would be the very best Scotch whiskey, with a little ice and plain water. There is something in that exquisite, smoky liquid that can't be imitated or duplicated, and that does things to my very soul. As it is, we drink Scotch only when it is given to us, and rely generally on rum and Kentucky bourbon. We have found it necessary and pleasant to get really tight, and in private, not more than three or four times in the last two years, and we must drink less than a third of what we drank in the years before. However, we should find it very difficult indeed to give up that last dollop or two of good liquor in the sideboard.

It seems to me right and natural that the very poor should cling to their liquor to the last ditch. Extreme poverty does things to one's deepest insides, and these in turn do things to one's view of humankind and the universe. It seems to me quite natural that Repeal was one of the first of the safety first measures taken by that most intelligent defense of an old deal, the New Deal.

In fact, I don't think that any of us can afford to look at nature and at the major facts of the human situation while dead sober. Then they seem serious, irrational, and rather dreadful. When one looks at them in a very mild state of intoxication, they seem equally irrational, but more strange and amusing than horrible. I suspect, furthermore, that the slightly drunken view is the true one, more true simply because perceived with greater sympathy. Organic nature is not, I think, completely sober, and we cannot be so if we wish to understand her, and smile. Our sober little minds are cold, and she is warm and large. It was a god, Dionysus, who gave us both wine and tragedy, which is the noblest wine of all, and which is never cold and dreadful. When we are forced to neglect this god, we do so at our own peril.

Interiors with figures.

In the winter, bathing is only a chore, involving most of the resources of the household, planned carefully in advance, and executed as quickly as possible. Before we were married, I found winter bathing also a Spartan test of character, because all the comforts and refinements have been introduced by Peggy. In fact, Peggy should have lived in ancient Rome, or in modern Hollywood, because, for her, bathing is almost a religious ceremony, enlisting all the senses in the service of the spirit, like the Mass. Here at Oak Hill, she is like a Roman Catholic isolated in a community of Quakers. Still, she is not lacking in ingenuity and persistence, and does surprisingly well by herself, while I, like a nostalgic Protestant at Chartres, wallow surreptitiously in what was not intended for me.

One winter, we bathed chiefly in front of the fire in the old soapstone stove in our bedroom. Using towel racks, towels, and rugs, Peggy made a sort of nest of warmth. Of course, one had to be careful not to scorch one side, but at least the other side did not shiver constantly at the same time. For fun, economy, and convenience, we used often to bathe at the same time. I often wished that I were a painter, and could record some of the comedy and sensual beauty of that scene. One felt it in terms of Daumier or Cruikshank, of Degas or Rembrandt.

Later, Peggy used one end of a back hall, above the kitchen stairs, to make what we call a bathroom. Here, behind a big screen, and next to a large window, were concentrated, for labor-saving and convenience, all the basins, water pitchers, waste water cans, wash tables, towel racks, medicine chest, dressing table, and mirrors. The electric range had supplanted the old coal one, so that a real stove could really heat both the room and a tank of water, with spigot, firmly fixed on top of it. Some of this heat managed to reach the bathroom, and was supplemented by two little kerosene stoves. If less romantic than the nest in front of the open fire, this arrangement was more convenient. And, as Lewis Mumford has remarked somewhere, there is no sound reason for a completely frosted bathroom window, cutting off the sight of whatever trees and sky are available. In the city, he suggests bathroom windows neck-high and clear, and in the country I recommend a very large window, quite clear, and curtained only against extremes of heat and cold.

Not that, in sub-zero temperatures lasting for weeks, this place is a temple of sensual delight, and not that, whenever we think we can, without wearing out our welcome, we do not avail ourselves of the generosity and bathrooms of our friends in town. It is hard to communicate, to anyone who has never lived without plumbing, or who is not so candid and ardent a sensualist as Peggy or myself, the actual ecstasy of a leisurely, elaborate tub in a perfectly appointed bathroom.

We shan't soon forget the night of one Charity Ball, in the Christmas season. The temperature was hovering around zero, and there was a heavy snow on the ground. Peggy and I had been working hard all day, and were very tired, very cold, and quite filthy. We had been invited to a dinner party before the ball. Just before James left in the truck, we took the precaution of seeing whether we could get our car out of the garage and onto the road. We got it out, and then very completely stuck. We got out the chains, ropes, flashlights, sacks, boards, and all the rest of that familiar paraphernalia, and labored for a while to no avail. It was getting late.

At that moment (may their tribe increase!) one of our friends the Rowans called us up and invited us to bathe and dress at their house. We accepted at once, called up Harold Prentice to come out and get our car if he could, some time during the evening, dumped our dress clothes into bags, all piled into the truck, and drove to town. We had at that time for a day or two, until she deserted us on the arrival of the guests for whom she was intended, an old darky cook, very cold and rather pathetic. Peggy drove, and piled the old woman in beside her, while James and I sat up back, with our backs to a wind that went through us like needles. I had my pumps in my hand, with my coat over an arm, the tails flying in the wind. We dropped the old woman at her corner and proceeded to the Rowans' whence James took the truck to his home.

All the Rowans had dressed, and were in a hurry, but characteristically had time to turn the bathroom and the best bedroom over to us, and to supply us quickly with tall, stiff drinks of hot whiskey. They are Roman Catholics, and hope, I suppose, for rewards hereafter. If anyone is rewarded, they will be. I hope that on some night when it is cold in Heaven, and their wings are frozen, someone will send us up to them from Hell, bearing hot water, towels, and hot drinks.

An hour or so later, we sat down to dinner, in our best bibs and tuckers, with Peggy looking and acting as though she had spent the day in the most refined and hedonistic leisure, and had just emerged from a fully staffed apartment on Park Avenue.

We have plenty of little experiences that could document a Marxian doctoral thesis on "Class consciousness in America as influenced by certain other elements, such as friendliness and a sense of the theater." For the author of such a hypothetical thesis I may add the fact that for us this dinner and ball were pleasant less for themselves than for their relation to what preceded and followed them. We had an extremely good time; if the transitions had been easier, from and to similar experiences, we might well have been bored to death.

Or perhaps these are simply items for the "Eating One's Cake and Having It Too Department."

Landscapes with figures.

In the summer we bathe, in our own peculiar ways, almost as easily and pleasantly as anyone. The bathroom is cool and dark, and not hard to manage. If we miss a shower bath or tub too much, all we have to do is to wait until James has gone home, and bathe on the lawn. We use both buckets and washtubs. I can't sit down effectively in a washtub, but Peggy can. (Incidentally, Peggy found an old child's tub in the attic. Her mother, when here on a visit, painted this white for the use of visiting children, but Peggy herself fits into it very neatly, without using a back-breaking amount of water.) One advantage of bathing on the lawn is that the water is so easy to get, and so easy to dispose of. Another is that one can splash and spill and pour with abandon. The most important is that it is such pure sensual fun to be naked and bathe in the open air, on the green grass, under the trees, with gentle airs, and slanting sunlight, or even moonlight, caressing our bodies.

It takes a little more time and effort, but is much more fun, to go down to the old swimming hole in North Fork, a creek a mile away in the valley. As a boy I used to walk down there and back, but driving is better, because one doesn't get hot all over again, climbing the hill on the way home. The hole we use is a fairly good one, over one's head at the deepest, with a moderately hard and clean bottom. There is a place to dive off, and there is a cleaner rock and gravel bar from which one can wade in. The current is strong enough to make the water seem fairly clean, whether it is or not. In a wet season, the water is too muddy for any use, and in a dry season we suspect it of infection. Still, we have been able to use it a good deal, with the water green and relatively clear, and we have not yet died of typhoid.

The hole is on a back road, and is fairly private, so that sometimes we can go in naked. There is a single-track railroad nearby, which makes us nervous about the dogs, but the trains are infrequent. We always take two or three dogs with us, for the fun of it, and to bathe them as well as ourselves. Jack insists on coming along, and doesn't mind swimming, but hasn't the enthusiasm for it of some of the cockers. Geoffrey is the most fun, for he has a small boy's passion for it, and loves to chase sticks. We had particular fun with Peggy's niece, who has been brought up on the surf of Cape Cod, but was far from scornful of the less sensational delights of our old swimming hole.

The swimming itself, of course, is nothing much, but one can at least be completely immersed and cooled, and come out feeling a new person. There is nothing like a long, deserted, sandy beach on the ocean, or even a steep rocky shore into the sea, with surf, salt water, and gulls, but our swimming hole provides more modest pleasures of its own. It is sheltered by willows and sycamores, and is visited, if one is quiet, by little green herons and other water birds whose names I shall never know. And one can lie on one's back in the water, in sunlight, or shade, or moonlight, look up at the silly little white clouds in the deep blue sky between the treetops, and listen to the distant, bucolic sounds of fields and barnyards.

More sensational, if more remote, is the waterfall and pool at the farm at Bellbridge. The creek is about seventy-five yards wide at this point, and the fall, at low water, is eight or ten feet high. The rock is limestone, carved into a multitude of treacherous and fascinating holes, caverns, and under-rock streams. The pool below the falls is wide, deep, and nearly always clean, in part, at least. It has a gravel bottom. The falls are always changing in volume, sound, clarity, and general appearance. We have seen that creek a deep, raging, muddy torrent, several hundred yards wide, wiping out farms and fences, and rushing along with such weight and savage power that the falls were only a ripple, throwing up a desperate wave. At very low water, there may be only a stream or two, a few feet wide, very clear and very gentle.

One summer a creosote manufacturer miles upstream so polluted the creek for weeks that the water could not be used by man or beast. All the fish were killed, and fouled the very air. A man who could do that should have been strangled at birth. He wasn't, and so there had to be committees, appeals, and protestations. The creek was finally clear once more, but there is nothing I know of to prevent a repetition. There is still a fair amount of fishing in these streams, for bass.

This creek is now our enemy, but oddly enough, the waterfall is one of the joys of our life. It has a remoteness and beauty, and a mysterious life of its own, that can not only take the weight and sting out of scheming, calculating, and worrying, but also make one almost forget the destructive power of the creek itself. When I tramp those fields, and go into those farm buildings, I always have some hope or fear, some scheme or worry, on my mind, but all I have to do is to walk down beside and over those falls to get away from all that, and sink into the most blessed relaxation and peace. That waterfall was there in the wilderness, long before there were any farms or houses within hundreds of miles, and it will be there, as remote, mysterious, and lovely as ever, long after the farm has been washed away and we are all dead.

It can't be said that this fall is an unmixed blessing for everyone concerned. It attracts picnickers, rowdies, and miscellaneous fools who are careless with fire and who leave papers and rubbish all over the place. Our hay barn has been burned down, and we can't get anywhere near enough insurance. Also, Mr. Kincaid, who is neater and more fastidious than many women, has spent many and many a Sunday evening and Monday morning cleaning up after imbeciles. If everyone were careful with fire and rubbish, we could leave the place open. As it is, we have to close it as tightly as possible, and Mr. Kincaid has to carry on an unceasing war against trespassers.

And Mr. Kincaid doesn't even have the compensation of swimming, wading, fishing, and idling himself. We have picnics up there, with our friends, from time to time, and not in summer only, but we are always careful to clean up, douse fires, and close gates, and he knows that we are. Quite aside from the stupid question of ownership, I don't think he minds our fun. Once when we went up there for swimming, with friends, the falls were impassable, and he suggested that we go downstream to an easier place, on his own property, very carefully marked "NO TRESPASSERS!" I shan't soon forget the two girls walking down that back road in their swimming suits, and encountering Mr. Kincaid, who was cutting alfalfa. He shoved his hat back on his head, and looked at them, quite without embarrassment, and with a friendly smile, but somehow in a way that suggested to me—perhaps quite erroneously—that he had never seen anything like that in his life. Why can't everyone have a little loafing on shady banks, and swimming in the sunlight, with good-looking women? I have been told that I always ask too much of life, and of human nature, but this doesn't seem to me to be asking very much.

Kilts and a beard.

Next to undressing, I suppose one of the greatest of personal and uneconomic pleasures might be dressing. As it is, we allow other people's interest in profits, and our own timid, conventional, and bogus ideas of dressing to spoil most of the fun. According to Eric Gill, who has thought about this matter to better effect than anyone else, including all the tailors and dressmakers in the world, men and women could dress to be as comfortable as they can in different kinds of weather, to be free and protected in different kinds of work, and to express their feelings of their own dignity and beauty. And I would emphasize a point that Gill makes only by implication: they could dress even more pleasurably and effectively to play their parts, not only typical and economic, but also individual and fantastic, or "bovaric."*

*See "Ad Astra," below.

Peggy and I compromise, like everyone else, but not because we like to fatten the stockholders of Best & Co., or because we are not critical of ideas and conventions; we do so simply because uncompromising individual dress reform costs more, in conspicuousness and isolation, than it is worth. Still, I chafe in more ways than one.

When we are at home and at work, in all seasons, we can let ourselves go, and we do so, with great pleasure. In winter, Peggy is apt to wear woolen underwear, ski-pants, several brightly colored sweaters, low shoes, galoshes, a leather jacket, a small felt hat with a small feather, and brightly colored mittens. I usually wear the new flexible breech clouts, a heavy shirt, a sweater, ski-pants, heavy shoes, galoshes, a football hood or a blue denim coat, a hunting cap, and loose cloth gloves from the five and ten. In the summer, Peggy wears sun-back dresses, or shorts, and enormous straw hats that make her look like an animated toadstool. One of her grandfathers was killed by the sun, so she fidgets when I go bareheaded, although I have heavy hair and hate a sweaty hat. When I can bathe on the lawn in the evening, I often wear shorts and leather moccasins; otherwise, blue denim pants, with a heavy belt, a light shirt, and heavy socks and shoes, are cleaner and more comfortable. Pants are unsatisfactory, because they get wet and caked with mud around the bottom, but any kind of boots are too hot and heavy. I don't wear overalls because the bibs are hot, and the straps bind the shoulders.

In short, work clothes are the easiest, cheapest, and most colorful. You can buy bandannas for a nickel apiece, heavy yet comfortable and wearing shoes for three or four dollars, and shirts for fifty cents. All these wear a long time, and when they go, they can be thrown away. Sometimes Peggy buys, by mail from New York, very smart and special work clothes, but these are frills, and we could both dress comfortably, efficiently, and appropriately for ten or twenty dollars a year, apiece.

When I go to town I usually wear my work clothes, because I can't be bothered to change, and because I'm proud of them. Only my town friends, and not my neighbors, consider this an affectation. It is easy to spot a farmer from his clothes, as well as from his face and hands, the shape of his back, and his walk. But I wish the costume were even more appropriate and distinct, like that of the railroad men, who wear their clothes with such pride and swagger. When I was a schoolmaster, I should have been glad to wear an academic gown, and I should be as proud, now, of a writer's clothes, if there were any, as I am of my farm clothes. Why, as Eric Gill asks, must all men, including kings and hodcarriers, try to dress like city clerks? Why do we leave all this fun to jockeys, railroad men, judges, and clerics? (And I would add, if we don't soon wear proudly costumes of our decent work, it may be that the Fascists will be trying to dress us all up like tin soldiers.)

It is when we dress up that we have to have money, or imagination, or both. Peggy has more choice, and can be more individual, but has to spend much more money; I have to conform to a dull and stupid pattern, but I can do so very cheaply.

Peggy's house dresses that I like best are cheap prints with tight bodices, short puffed sleeves, and enormous skirts. She is now making these for herself. In town, and on little journeys, she is apt to wear sweaters and skirts, or severe little suits, or sleeveless dresses. She says, and I think rightly, that most women in Chillicothe, according to Eastern habits and standards, overdress extremely, and sometimes even succeed, against her will, in shaming her. She rarely gets a chance to wear evening gowns, and when she does, revels in it. In short, she dresses with good sense, individuality, simplicity, and style. In order to do so, she has to take great pains, which is perhaps inevitable, but she also has to pay through the nose, which seems to me more open to question. I should like very much to know why simplicity and good taste, in women's clothes, cost so much more than their opposites. I have an idea why. And I also have an idea that if clothes were produced for use, the popular taste would soon prove to be extremely good.

When I have to change a little from my work clothes, or feel like it, I refuse to rush into a clerk's costume, and compromise on the tweeds, flannels, and so on that are one of the achievements of English and eastern American schools and universities. I am rather messy at best, so it probably doesn't become me to suggest that the young business men hereabouts dress timidly, uncomfortably, and with too obvious care and conventionality.

I can say, however, that if there is anything more false and pathetic, in dress, than myself at a dinner party, or my farmer friends at a meeting or funeral, I don't know what it is. Why are we so afraid of each other and of our wives? Why are we so ashamed of our bodies, and of our jobs, which make stiff collars and waistcoats, on any occasion, so irksome and false? Can't there be, some day, a dress costume that will give the American farmer the naturalness, comfort, dignity, and pride that are his by rights? Even if such a costume should appear, in the next hundred years, it is quite impossible to imagine it now.

What I'd like to see on dressed-up farmers and wear myself is kilts, with soft, loose, low-collared shirts, and very short jackets. And I'd like to see beards, and wear one. Why must we go through this bore of scraping our faces? Only because our wives want us to? If so, why do they? Are they ashamed and afraid of our characters and work, so that they want to make us look like town gallants? A clean-shaven face may be appropriate for a modernistic apartment or office, where everything else is clean-shaven and sexless, but it is not appropriate on a farm. A razor is as timid and citified a thing as a girdle to flatten buttocks whose glory is their curves, or a brassiere to lift up breasts that can be lifted in health and pride, or a pair of pants with buttons or—nadir reached by man in the machine age—a zipper, instead of a codpiece.

"For this relief, much thanks."

For our grate fires we find a combination of coal and wood the most effective fuel: wood for starting fires, or holding them, or for taking the chill off a room of a cool spring or autumn evening, and coal for heat. Anthracite is an expensive rarity, and we use a bituminous called semi-Wellston that comes, now, for about six dollars and a half a ton.

One can buy coal of uncertain quality, for three, or three-fifty, from men who bring it here from neighboring counties to the south and east. These men have small mines on their own farms, and mine, haul, and sell the coal themselves. A few have regular customers, but mostly they appear in town with a load of coal and dispose of it for whatever price they can get, to whomever they can find. I have had a man with a load stop me on Main Street and try to sell it to me then and there. If you know good and bad coal when you see it, and shop around long enough before you run out, you can sometimes pick up bargains. Once, in such a deal, I failed to distinguish between two cousins who had the same name, claimed ownership of the same mine, and were on the point of shooting each other. When it looked as though they might shoot me instead, I decided to stick to the regular dealers.

The only compensation that we have for the loss of our oaks, aside from pieces of silver, is a huge supply of firewood. For two years now, in autumn and winter, James and I have spent any time we had left from other things in working up this wood, and it looks as though we'd be doing the same thing for the next twenty-five years. In time, we may have to stop buying coal entirely, and spend more than spare time with saw and ax. An accidental dividend from this old wood may have been an attack of termites on the underpinnings of the house, repulsed at a tidy cost of $140. The termites might have been in the house for years, but it might have been cheaper if we'd burned all our wood in the fields, and bought cordwood, which in turn might have brought in the termites. And how many brains does it take to foresee such things?

Once I succeed in putting aside little reflections like these, and once I have grown accustomed to the ugly carcasses of great trees that I have loved, I thoroughly enjoy sawing, chopping, and splitting—which I call hewing, a term I never hear. They require skill and knowledge that can be acquired by a city-bred man only very slowly. They tax your strength to the limit, yet you can always stop and rest. The results, though meager considering the work involved, are definite and indisputable, as in painting. When the leaves are falling, and the winds are rising, and the snow comes, a neat and sizable pile of firewood is a Good Thing. When I find myself getting nervous and touchy, and all my little worries and complexes are beginning to warm up and hop about, a long afternoon in the wind and cold, with ax and saw, is a sovereign cure. For the good of mankind, dictators should be required to take up the ax and saw before they come to power, instead of after they are deposed or shot. Carlylean interpreters of history should note that the Kaiser is sawing wood in Holland afterwards, while Stalin did his stint in Siberia first.

When I came out here as a boy, an old Negro, Hubert's father, began to teach me how to use an ax, and I am better at it than some people I have seen, but it will be years before I can always hit the right place, with just the right force, and always get my whole body into it, without any straining and wasted effort, and with the grace of a good man with an ax. I have learned

how to sharpen an ax, for splitting and chopping, and I have
learned that for keeping an ax head on, a piece of pulpy poplar
wood is better than any metal wedge. I have at last learned how
to file and set a saw, not really well, but well enough to get along
with the work. It will be years, if ever, before I learn all the kinds
of wood, and know what each is good for, and how it should be
handled. But I have seen enough old carpenters and woodsmen
go wrong, not to be quite discouraged.

Another thing I like about hewing wood is the pleasure of
working outdoors, in winter, with James. We have always been
able to swing a crosscut saw together, rhythmically, without
wearing each other out, and that's more than some teams can
say. James works along very steadily, and yet even he will admit,
now and then, a little weariness, and is not sorry when I have
to call a halt. He can be silent pleasantly for hours on end, yet
sometimes he feels like talking, and when he does, he has something to say. He is always alert to the condition of sky, wind,
snow, and air, can tell the different kinds of hawk apart at great
distances, and can identify almost any track in the snow. His
response to the scene about us is never wholly practical. He
likes to stop for a moment and watch the dogs at play, or hot
after a rabbit, as well as I do, and I can see in his eyes, I think,
that a bird on the wing means more than something to be shot
or spared, for some reason. I like also to draw out, if I can, his
superstitions about the moon, or about animals, or whatever.
He has plenty, but he wears them lightly, and I am sure that
most of them have in them, somewhere, a germ of truth. Sometimes, gingerly, I like to try to explore his knowledge and opinions. In a few ways, at least, this Negro who has raised a large
family, and has never been outside Ross County, except to
Columbus, is closer to me than some people of my own color,
age, and class, whom I have known all my life.

One day, I remember, we were both kneeling in the snow

and mud, on opposite sides of a huge log, using a crosscut saw. We were near the end of the cut, but I had to stop for a minute to get a fresh start. I was too tired even to stand up and stretch. Somehow we got to talking about war. It appeared that in the last war, because of his family and occupation, James had just escaped the draft. I dared to ask him whether he would sign up, or help in any way, in another war.

"No sir, I wouldn't. Not me."

This was more definite than anything I had ever heard from him, on anything.

"Why not?" I asked.

"Well, sir," he said, "I may be wrong, and I ain't read much, but I look at it this way: I don't see that anyone gets anything out of it, one way or another, I don't want to kill no one, and I don't want to get killed."

The best time, of course, is when a darkness and a new chill have fallen on us, and we guess at the time, and pick up the tools, and start back to the house—James to his carrying of wood, coal, and water, and to his dishwashing, and I to my feeding and milking. When we get through we are tired, but not too tired, and just cold, wet, and dirty enough to look forward to warmth, dryness, hot water, rest, and food. Sometimes Peggy comes out to meet us, and the dogs all spot her and run to her, barking with delight.

A couple of hours later, the work is all done, and James has gone home, and I have bathed and changed my clothes, and Peggy has supper ready before the fire, but has taken time to make us both a little drink. So we sit there, a minute, before the fire, and raise our glasses to each other, and talk about this and that, and fall silent. Then we eat, well, and clean up the table. Sometimes we call up Gerald Rowan, or somebody, and go to a movie. Or else we stay at home, and listen to music, or play chess, or talk, or read, sometimes aloud, some book that has

blood in it, some book whose "iron English rings on the tongue."
Then we put the dogs in their kennel, take the air a few minutes in the cold moonlight, listening to a hound somewhere far
away, and go to bed. And after a while, when we are lying close
to each other, and the windows are open, we watch the black
limbs moving, and listen to the wind, or to a mouse in the wainscoting, and watch the moving firelight on the ceiling.

If there is a god, anywhere, who has done this to me, let
him hear my thanks.

The society of trees.

There is also a society of trees, which is certainly more important to human society, and is more important to me personally, than the gracious society of flowers. When your old trees,
your old friends, are dying, and being hewn into firewood, not
only a reasonable attention to the first principles of agriculture,
but also an inner hunger, a psychological necessity, demands
their replacement. They cannot really be replaced, within a
century or two, but as any old person, and especially any old
person without children and grandchildren, will tell you, a beginning is already something.

However that may be, in her later years my aunt set out
scores of young trees, transplanted from the wooded cliff, and
bought from nurseries with money that she needed for her own
comfort. We have been more attentive to our comfort, but we
have carried on, and in no other expenditure of money, time,
and effort have we got greater satisfaction.

Our first spring, we set out a number, but we didn't get them
tall enough, or protect them properly, and the cattle killed all
we set out in the pasture. We cut a score of willows and stuck
them in the dampest places, but only two, including a weeping
willow given us by a friend, have survived. That fall, the outer

parts of the lawn were not kept cut, so there survived a number of young sassafras trees, now from six to ten feet high. For my birthday that autumn, Peggy gave me two fine young Austrian pines, about five feet tall. To these we added three others, and fenced them all with barbed wire, but calves and the drought got in some dirty work. We have agreed to give each other trees only—or at least trees always—for our birthdays. At the same time, I bought three Lombardy poplars, which I like at a distance, silhouetted against the sky, partly because they remind me of France. These were attacked by insects, but two have survived, and in time I may get cuttings from them. The following spring, we set out about fifteen trees, from the cliff, from the farm at Bellbridge, and from nurseries, but the drought killed almost all of them, after they had started well. I carried and hauled such water as I could, but it was not nearly enough. We decided to set out trees only in the autumn. That next autumn I bought two good hemlocks, and some smaller evergreens, from a man who had hauled them from North Carolina, and these all seem to be doing well. We also transplanted half a dozen sycamores to damp and open places in the pasture. Some farmers find sycamores a nuisance, but to me they are among the best of the trees in this part of the country. They are to us what birches are to New Englanders.

Tree surgeons keep coming up here hopefully, but I have never felt that I could afford to hire any of them. It would cost me thousands of dollars merely to get them started, and I am not wholly convinced of the efficacy of their work.

During my first summer here alone, I made a chart of the remaining old trees, and of the young ones set out by my aunt. Not counting the wooded cliff, they added up to only about fifty, of which perhaps a third were not more than ten or fifteen years old. Of the great white oaks that gave the place its name we only have six left alive and in moderately good condition.

The others include black walnut, catalpa, red cedar, redbud, sassafras, hemlock, crabapple, red birch, Norway maple, dogwood, black ash, tulip, pin oak, green ash, sycamore, shellbark hickory, yellow locust, and white elm. If I were a poet, I'd write a poem to each kind, as represented here. I might imitate the translations from the Chinese, and give the whole series a nostalgic, elegiac tone. This form and manner might be altogether too appropriate, because they say that unless reforestation and general conservation become a national passion, this country will soon become as desolate as China.*

But despite a public attitude that until recently has been casual and destructive, our country is still glorious with trees, in spring or summer or autumn or winter; in sunlight, or cloud, or moonlight, or fog; at dawn or at dusk or at high noon; in dead calm, or in gentle airs, or even, while you hold your breath for them, in thunder, lightning, and hurricane. There is still a great wealth, to be guarded, and added to, but as always, I can't say very much for personally owning any of it. Who can own a tree?

This reminds me of a great wealth just beyond our meadows to the south: the large orchard of apple, peach, and cherry trees "owned" by our good neighbor Ralph Stone. At least, it is Mr. Stone who does all the work and makes all the money, or takes all the losses. I don't imagine that it has ever occurred to him that he might very reasonably charge us rent for this orchard of his, but it may be that we get more real pleasure out of it than anyone else, his customers, his family, and himself included. With his connivance, we filch a few fallen apples, but that isn't what I mean. It makes me glad merely to think of that place. We walk through it at all seasons of the year, and know many of its colors and smells and shapes. Our dogs also are

*The Chinese recently sent a large contribution to our Red Cross, for flood refugees.

greatly in debt to Mr. Stone, or, if you prefer, to God Almighty. (Of Whom I am reminded by thinking of a time when I worked in that orchard as a boy, picking apples. Another boy, colored, hearing thunder, said: "Dat's de Ol Man, rollin his bones." An old colored man stuck his head out of the top of a tree and said passionately: "Don you say dat, boy! Dat's de Lawd God Almighty!") At the other end of the orchard you reach the south brow of the hill, above a wide and beautiful valley, beyond which there are, first, five hills. Once, when I was less of a house-holder and farmer, I had time and fancy to name these five hills Geoffrey, William, Thomas, Anton, and Marcel. Figure that one out, if you like. And anyone who isn't stupid enough to spend all his time trying to farm, or to make money, or books, or something, knows that a neighbor's orchard is a perfect place for making love.

Coming, a real job.

In all this talk about gardening, eating, drinking, bathing, wood-chopping, tree-planting, and loafing in a neighbor's or-chard, I have, "just like a man," dodged as long as may be the drudgery that can never be escaped by the farmer's wife. To be sure, Peggy is relatively lucky in the help that she gets from James, from an occasional girl in the kitchen in the summer, and even from myself, and whenever she feels imposed on, cab-bined, cribbed, and confined, all she has to do is to visit any of our farm neighbors.

For instance, in the winter, of a morning, James builds the fires, carpet sweeps, dusts, and mops the dining room, fills three kerosene tanks, empties water from and carries it to the kitchen and bathroom, brushes out the kitchen, pantry, and kitchen steps, sweeps the front steps and walks if it is snowy, and washes the breakfast dishes. At noon he stokes the fires and replenishes coal, wood, and water. In the evening he washes luncheon dishes, stokes fires, etc. When I am deep in a job of writing or something, he also takes on some or all of my chores. He also has regular days for scrubbing, vacuuming, polishing, and so on. In the summer, he is sometimes relieved of the dishwashing and some other inside jobs, and unless we are away, he never comes on Sunday. If we have a girl in the summer, she takes over the dishwashing, much of the indoor work, and as much of the cooking as Peggy can train her to.

Still, any housekeeper will see that with the organization of all this, the main burden of the cooking, the shopping, bed-making, dusting, darning, laundry-checking, and so on, ad infinitum, not to mention the constant care of from four to eight dogs, Peggy's job cannot be called a slouch. Yet she gets in a good deal of reading, painting, repairing, entertaining, and Little Theater work, and endless tramping about and horseplay with me. She does a fair amount of volunteer work, all over southern Ohio, for the Girl Scouts, and for a fortnight or so last January worked a couple of hours almost every day for the Red Cross, which fed, bedded, nursed, and entertained several hundred flood refugees. In addition, she is an active and helpful partner in my small farming operations, and when I am a slave to this typewriter, or away on business, she has a pretty good idea of what James is, can be, and should be doing outdoors.

I have already mentioned some of the repairs and improvements. The most important of the latter, so far, has been the introduction of electricity, with which we light, refrigerate, cook, mix, vacuum, churn, and do a little ironing. We got the current up the hill by guaranteeing to pay a minimum monthly bill of six dollars, for four years. We thought this was highway robbery, but there seemed little prospect of the T.V.A. or Farm Bureau lines coming this way for many years. Our bills have been the minimum in the winter, and up to nine dollars in the summer. Our wiring job, with floor plugs only, cost us $203.52, and when I found a neighbor's sons wiring their father's house, I was properly shamed.

Another element that lightens the drudgery of housekeeping in the country is the comedy of the thing. I am thinking of one day when I had my brooder in the kitchen. Peggy came downstairs and thought the chicks were making a lot of noise, so she opened the lid to see what the matter was. Out flew a hundred chicks, and the kitchen was filled with cheepings and the flurry of wings. I was outside, and luckily all the dogs were with me. Peggy laughed till she was weak and then pulled herself together and chased and lured chickens from the stove, scrap basket, coal buckets, galoshes, window sills, and everywhere else, until she had caught them all and got them back into the brooder. That afternoon, I may say, I "decided" it was warm enough to put the chicks out in their house, with a couple of kerosene stoves.

Finally, Peggy has a passion for moving furniture and pictures, and this is given free play by our semi-annual move into one half of the house and then out over the whole house again. What with this, and the doing over of the floors, there isn't a stick of furniture in the house that Hubert and I, and now James and I, haven't moved from room to room at least six times.

Now, as you may have gathered, repairs, improvements, and moving do not, while they are being done, please me one whit more than they please most men. But it isn't just the prospect of greater comfort for both of us that makes me endure all this, and mind my manners, while it is going on. I am also kept in my place by the notion that if anyone is willing to endure the horrible drudgery of keeping a farm home clean, orderly, efficient, and unobtrusive, she has a perfect right to almost any amount of improving and rearranging, because it is all this that gives the whole effort meaning, and changes it, in part, from drudgery to creation, to a real job.

In part only. Despite Peggy's evident pleasure in the remaking of this place, as our home, and in all the rest of it, I should be glad if we could earn enough money to afford a regular girl in the kitchen, and to shove all this a little farther into the background. I don't happen to think that household work can be a real job for a real woman, unless her major talents lie in that direction, and unless she has been trained, also, to make a satisfying job of it.* Furthermore, despite my admiration, for instance, for the ordinary Frenchwoman's making of a home, I do not think that this job, even when limited to direction, and even when including the intelligent raising of children, and all the best psychological adjustments, so to speak, can be a completely developing and satisfying job for a real woman. We are luckier in all these respects, and in most others, than most people, but Peggy is still restless, and even though she has much more nervous energy than physical strength, I shall be glad when she finds a real job, a real "specialty," as I have called it, for herself alone. I hope it will be as satisfying and exciting as mine, and much more profitable. In any case, I shouldn't notice it if the housekeeping job were half as well done, and I shouldn't mind it if it were a tenth as well done. And I am glad to take the risk of saying this in print!

*I understand that the W.P.A. has undertaken such training.

VII. People

A note on American restlessness and loneliness.

The restlessness that I have just suggested, in relatively mild strength, in a relatively intelligent and lucky woman, seems to me much more marked and dangerous in many other American women, including many with children. It is also notable in many men, and that is why it seems to me to go beyond the question of the position of women in homes and in society. Let's assume, for instance, that Peggy is able to master and neglect her incidental jobs here, and to look about for another, more personal, satisfying, and possibly remunerative. She will then be roughly in the position of most men I know, including myself. She will have to find something that both pays and seems worth doing, in itself. Now I want to suggest that this problem, quite aside from the constant and universal danger of unemployment, is more formidable than it has been for twenty-five years, and will probably remain formidable for another fifty, at least. Almost every man I know is in an economic jam of one kind or another. Ninety per cent of them are immature, take a short view only, adjust themselves to reality on this basis only, and either work themselves to death, or play the unconscious fool to forces beyond their comprehension, or do both at once. Ten per cent are mature, understand in part the forces at play, can take a long view as well as a short, and will survive, God willing, but are profoundly restless and unsettled. Of these, only a few professional men, free spirits, and lucky exceptions earn their livings and are happy in their work and proud of it. I have a fairly wide and varied acquaintance, but I can count on the fingers of one hand the men and women who have found work, and a kind of life, in which they can grow to their limits and do their best by both society and themselves. Perhaps it was ever thus. I doubt it. Perhaps it will always be thus. Again I doubt it.

This fact, if it is one, may have something to do with what I call the loneliness of mature Americans. It does not fully account for it, because this is a huge continent, it has been settled quickly by a wide diversity of peoples, and from what I have read I have got the idea that mature Americans have always been in a small minority, and have always been lonely. (Not long ago Clifton Fadiman wrote something to the effect that most of our great men have been, when alive, lonely "failures," like Thorstein Veblen.) Still, I wonder whether this loneliness is not now intensified, for all the improvement in transport and communication, by the profound changes in economic, political, and social relations, and therefore in ultimate attitudes and hopes. Of the 10 per cent I have mentioned, very few are able to see anything of the people closest to them in spirit, fewer still have adequate certainties for themselves, and fewer still are able to agree on the fundamentals.

These speculations demand qualifying and elaboration, and in this abbreviated form they are perhaps inconsistent, but they may serve as an introduction and background to the following sections on our own social relations. Since I use them as such, I must make it quite clear that I do not think that we ourselves are especially mature, or superior to our neighbors and friends, or that we are in any sense great, or even significant, failures. These preliminary speculations derive whatever relevance they have only from the fact that our personal histories, attitudes, and characters isolate us almost as much, and in much the same ways, as would the exceptional maturity, knowledge, and significance that we certainly do not claim. This will, I think, become clearer as I go on, and other actual or potential immigrants from one kind of life to another may perceive bearings of these remarks on themselves.

We are certainly restless in our economic insecurity or false and unjust security, though we think we are going in one right direction, in so far as we can do so alone.

More important at this point in the report is our unquestioned though not at all serious loneliness. I use this word more for Peggy and for others of characters like ours, in similar situations, than for myself. I happen to have spent a relatively large proportion of my time alone, and have therefore developed an immunity to a situation that might well make others lonely indeed. For whether we are lonely or not, we are isolated, and the following sections have to be introduced by a general confession of failure. I shall emphasize our minor successes, because they are more interesting, but their limits should be noted, and their small number, for a period of three years, should be remembered.

We have some animal friends, a servant who is yet a friend, a tenant who is yet a friend, a few good but remote neighbors, a few older friends, a group of companions in amateur theatricals, three or four side-kicks, and a family who take us in. Unwillingly, but inevitably, we are constantly baffling, frightening, and offending all sorts of people. Our pasts, and our strange purposes, isolate us from almost everyone. There is no one within seven hundred miles who understands, even with conscious effort, more than 75 per cent, at most, of either of us. No one really speaks my language, and despite her much greater ability in communication, Peggy remains a very small person in the center of an enormous field of very alien corn. We are aliens for life. If we were both killed in an automobile accident, not more than a dozen people, here, would be really moved.

Yet, as these pages may show, we are not quite alone; our failure in this respect is not quite complete. There is warmth in us, and warmth in other people. Almost often enough there flows that current, simple, human, yet quite beyond conscious generation, or explanation, or control, that is a first necessity of any life, of any society, of any civilization, and that is one of the things that keep Peggy and me, and everyone else, from being quite solitary, dreaming, and raving (if happy) lunatics, with automatic pistols in our pockets.

The good companions.

If we were limited, for company, to human beings, we might not be living here at all. I have already tried to suggest that without indulging in vague and sugary nonsense, one can find personality of a sort, and company, in flowers and trees. I don't doubt that relations of this kind could also be found with snakes, birds, rabbits, squirrels, and all the rest, but the sad fact is that fairly conventional farm life, like ours, is conducive neither to friendships of this kind nor to true, fresh, and intimate knowledge of wild life, and of the wildness left in domesticated animals. Of course, good farmers are apt to know more about predatory and game birds and animals than most city people do, and their knowledge is not always defensive or offensive, but as Thoreau pointed out, the domestic service of animals is not helpful to disinterested contemplation of nature. I have not been surprised to find that it is the poorer farmers, and the drifters, whose eyes and minds are more open to these wonders that do not so directly concern us.

Among the animals, our dogs are our closest companions.

Horses have a nobility of their own, and for me the suave egotism and independence of cats have a special charm, but there can't be any question that dogs are the best companions. I find the non-human remnants in dogs—most apparent in relation to sex, food, and scent—among the most interesting things about them, and I can't raise a cheer for their extreme dependence on men, but no one has better cause than I to be grateful for their friendship. Our four dogs have, like all dogs, individual personalities, and these four persons are almost as important in our lives here as if they were our children. They demand, as always, more than passing attention, and anyone who is bored by talk about dogs and children can skip.

Jack, nine years old, is, as I may have said, a sable and white collie, with a strain of what they call, in this region, shepherd. He is larger than most purebred collies, his head is chunkier, though well formed, and he has more intelligence. Because of her bad experience with other dogs, and because of her solitude, my aunt kept him tied most of the time, though she always let me give him runs. He is very sweet-tempered, fastidious, and sensitive, often his spirits are high, but sometimes he succumbs to a certain proud melancholy.

He is a living disproof of the adage about not teaching an old dog new tricks. After the death of his mistress, and even after his year in New York, while I was in Connecticut, he attached himself firmly to me. He never did like the Brookses. On Peggy's arrival, he went through perfect orgies of jealousy and self-pity. After a few weeks, he adjusted himself, and now he is quite as much attached to Peggy as to me. More remarkable was his adjustment to our cockers. He had had this place entirely to himself, and hardly had he accepted a new mistress before his domain was invaded by three new dogs. He insists, gently but firmly, on his share of attention, and sometimes, when three or four dogs are trying to hang on to his ruff, he politely dodges them, but he has never been actively jealous. He graciously accepts the hero worship of the smaller dogs, and sometimes plays with them for hours on end. Often, when he makes a dash for a squirrel, or a rabbit, or a turkey buzzard, he looks over his shoulder to see whether his gang is following him. And in their encounters with other dogs he always rushes to their protection. I have seen him fight savagely when attacked, and sometimes without that reason, but any kind of fighting on the place upsets him, and he will not tolerate any fights between the cockers. When Geoffrey and Peter begin growling at each other, and circling each other, stiff-legged, Jack steps between them, lowers his nose to theirs, raises one forepaw, and gives just a hint of a growl. There he remains, immobile, until the cockers give up and try to act nonchalant, by scratching themselves or admiring the view.

Jack's impetuousness with livestock, including chickens, is annoying, but his only major sin is stealing food. Once we were about to go on a picnic with guests and discovered that Jack had eaten all the seventeen hamburgs we had ready to broil. We rarely put him on a leash or chain, but he doesn't go away for more than an hour or two, unless he is after a bitch in heat. One time he kept visiting a stray bitch that had stopped at a neighbor's house, until the neighbor got tired of both of them, and shot the bitch. But Jack found himself another, and stayed away for two days and nights. I drove and tramped miles across the countryside, looking for him, asking everyone I met, and thinking of barbed wire, steel traps, sheep, and shotguns. It was literally sickening. Finally he ambled home, with his girl. Peggy grabbed him, and Geoffrey took that opportunity to run off with the lady. However, he wasn't up to her speed; in a couple of hours she discarded him, and he came home, looking rather crestfallen. One summer morning, soon after she came here to work, our hired girl of a summer met Jack on the pike. Two men had him by the collar, and were trying to make him get into a car, but with real courage, and in spite of threats, she rescued him and brought him home. He is in excellent health, and I hope will live to be a very old dog. He is not only a real character, and a friend whose devotion is almost embarrassing; he is a living and irreplaceable link between my aunt and ourselves.

Two of our present cockers, and Peggy's first dog here, were given us by a man I worked for at one time in New York, named, shall we say, Pickering. Mr. Pickering is a coal merchant, gentleman, sportsman, collector, philanthropist, and whatnot, who happens to be a strong and fascinating character in himself. I'd like to write a sketch of him here, but this is really no place for it, and besides, the man arouses in me such deep and contradictory emotions that I probably couldn't attain much objectivity. One of the emotions is the most profound gratitude for, among other things, about as nice a pair of small friends as Peggy and I shall probably ever have.

When Mr. Pickering sent us our first two cockers, Jeanne and Jerry, they arrived in such a state of terror that it took them weeks to recover. Jeanne may have been frightened badly before that journey, because she has never fully recovered. Jerry was a gay and completely charming little dog, utterly devoted to Peggy, and very pleasant to everyone else.

Jeanne is a little blue roan, with black mitts and stockings. She has English cocker blood, which is manifest, but she is much smaller than most English cockers, and I think the English head is better on a small dog. (For several reasons, we signed the petition to the Kennel Club opposing the crossing of English and American cockers.) She has a rather decadent loveliness, and her ears, her mitts, her expression, her sweetness, and her timidity strongly suggest a Victorian lady long immured by a domestic tyrant. But she has a splendid nose, and a passion for any game, especially feathered. A good half of this Victorian lady is a most swift and efficient killer, and when we find dead chickens, we have a very good idea who killed them, and this, alas, means beating the Victorian lady sitting there so delicately in the library. You may know Jekyll-Hydes, but have you a little Elizabeth Barrett Borgia in your home? And Mrs. Browning, too. Jeanne was an excellent mother, but it didn't alter her figure, and she remains an exquisite spinster.

After Jerry's death, we didn't know that Mr. Pickering was going to give us another male, so we bought a male pup, especially for Peggy, from a breeder in a nearby town. This dog had good ancestry, and at eight weeks had very promising marking, conformation, and disposition. We knew we were taking a chance, but Peggy wanted to raise her own dog from weaning. He was a black dog, with a few good white markings. Within a few weeks he turned into a perfect caricature of a cocker, with a large setter head, a foul-marked nose, elephant ears, and short bowlegs. When his breeder came to see him, she was dismayed, and wanted to give us in addition a good dog, that is, from the show point of view. But we didn't accept, because we had got our money's worth, and had enough little mouths to feed. Except in appearance, Geoffrey is an excellent dog. Peggy has fed him well, and he is as tough as hickory. He is also smart, friendly, cheerful, sensitive, jealous, temperamental, and mischievous. He is a confirmed Peter Pan, and with Jack striking noble poses, and the other cockers such obvious blue bloods, Geoffrey supplies a certain necessary plebeian touch. His adventures and misadventures would make a gigantic picaresque novel, which I shall not write here.

Peter's registration name is something like this: Canterbury Treble-Bob Major, A.K.C. 95047862. This isn't his real name and number, but I don't want to embarrass him by publicity, from which he would shrink in silent pain. He also is a black and white, but very handsome and aristocratic-looking, and he is as gentlemanly inside as out. He is affectionate, intelligent, and invariably courteous, with the other dogs, with us, and with strangers. When he arrived, he was ridiculously citified, but a few weeks changed that. Now he is a passionate hunter and when he wants to be, a happy roughneck. He was rather reserved until he got so much attention when we first showed him, and that brought him out wonderfully.

Jeanne's and Peter's litter of pups make another whole story of character, psychology, and canine relations, but I have to stop somewhere. I may say that raising a litter is very arduous, and extraordinarily good fun. It must be obvious that four good dogs are more than four times better than one, in almost every way. Still, aside from raising dogs for profit, when we should have to take great care not to become too attached to them all, I may say that we find four just about enough. When I was last in New York, Mr. Pickering wanted to give me a greyhound, but I had regretfully to decline. We may have to starve with four friends as it is.

Sometimes when Peggy and I are tramping across our fields with from four to seven dogs, and all of us well fed, healthy, and happy, enjoying each other and everything else, I see suddenly in my mind's eye a room or two in a slum, or in a shack in the hills, with a man and woman and seven children, not so well fed, not so healthy, and not so delighted by each other and by everything else. There is more than one thing to make a man absentminded, more than one thing that can come between him and the immediate experience of reality, which, I think, is the only good beyond question.

Old family retainers.

The name Hubert Brooks has already appeared in these pages. During the last seven years of my aunt's life, this Negro was her hired man. Hubert's father had worked here from time to time before, and still did the plowing and hauling, while

Hubert's mother occasionally helped in the kitchen, and Hubert's brother-in-law cracked the corn for feed. I think some other Brooks did the laundry. At any rate, my aunt used to remark that in one way or another, the Brookses got most of her money.

Hubert always tried hard, but he was incurably childish. He indulged in prodigious fantasies, and when these were not borne out by the facts, he grew sulky. He had the most astonishing powers of rationalization and self-excuse. My aunt understood him, directed his every move, and when he needed it, called him down sharply. Also, living alone, she told him many things about her personal affairs that he would not otherwise have known. She endured him as best she could, and even liked him, and he was sincerely devoted to her.

When my aunt died, Hubert was naturally quite upset, and began his endless speeches about his and his family's devotion to my family. My sister and I hired him to carry on the work, and with his wife and four young children, he moved into a limited part of the house, which we dared not leave uninhabited. We got a very capable friend of ours to pay him his wages and keep him out of trouble. This young man was most efficient and responsible, and so he and Hubert did not get along well. During our absence, and with this eagle eye over him, Hubert did moderately well, though I am inclined to think now that that winter of relative freedom was his ruin.

When I moved out, I was much occupied with writing a novel, and let the Brookses go their own way. I knew I was spoiling them, but I was unable to pay the price, in mental wear and tear, of treating them properly. Hubert's wife, Minnie, cooked for me, after a fashion, without extra charge. She was a cheerful soul, and worked hard, in complete chaos. When the Brookses finally moved to town, the kitchen looked like one in a slum.

One reason I was a bad employer was that I found these

people endlessly amusing. Working with Hubert, I could listen to his fantasies by the hour. The oldest boy was a solemn child named Clarence, who took much of the care of the younger ones. Then came twin girls, Myra and Mabel, whom I never learned to tell apart. They were very shy, and full of giggles. The youngest, and the darling of the entire Brooks family, was a little boy named Selwyn, who was superlatively cute, but was always playing with tools and losing them, and was always either underfoot or standing on the oven door, getting his pants dry. Once Hubert was upstairs, moving a wardrobe about eight feet high and several hundred pounds in weight. It toppled over on Selwyn's head, knocked him cold, and was itself demolished. Selwyn was brought to, and then fell asleep, and then, as though by magic, scenting death, innumerable Brookses appeared in the kitchen. After a couple of hours, Selwyn awoke, quite all right. The wardrobe had to be completely rebuilt. Despite my fuming, he used to play with the ax, until finally one summer evening he almost cut his leg off. Hubert protested that he had fallen on a broken bottle. In a few weeks he was quite himself again.

When Peggy appeared, a new era began. We paid Minnie four dollars a week, and Peggy took her in hand. They got along well enough, but Hubert of course resented Peggy from the start, since she didn't have any reason for being sentimental about either of them. Still, she spoiled them almost as much as I did. Everyone was relieved when they decided to commute from town in the old Model T sedan that ran only by the grace of God. Hubert's father died about this time, and Peggy and I went to the funeral. Old Mr. Brooks was a good man, and I'm glad he didn't live to the end of this story. When Peggy went to Cape Cod, the Brookses relaxed, quite unpleasantly, in a way hard to describe except by saying that I kept catching Minnie doing things like putting new cream in a dirty churn.

One Saturday, a week before Peggy's return, with her father

as a guest, I arranged with my doctor, for the third and I hoped last time, to have a slight operation in the next few days. I was fairly weak at that time. That night I went with a friend to a movie. When I got back to the car, I found a note from Hubert, asking me to stop at his house on the way home. He came running down from the porch, and told me he was leaving the job. Did I want him to come back to milk the next day, or to stay until Wednesday night? I chose the latter, and tried to make the parting as pleasant as possible. However, when, the next day, they refused to stay until Peggy could get another cook for her father's visit, and refused to help me get another man to see me through my operation, I sent them packing. As they drove off the place in their old Ford, for the last time, Minnie looked out of the back window with a gesture that had the brief and painful appearance of a thumbing of the nose.

It was no shattering surprise, but it was something of a shock. That man and I had worked hard together, and had rather deep memories in common, extending over both our lives. I had tried, at some cost to Peggy and to myself, to make every allowance, and to be as decent as I could. I imagine Hubert, in his way, had tried to do the same to me. Besides that gesture of Minnie's, I found, in the mess of the kitchen, another little souvenir: a ticket in a numbers game.

We have only heard rumors from the Brookses, and these have not been good. Sentimentally enough, I still feel somewhat responsible for those people. I sometimes wonder whether Selwyn is still playing with axes. But no more cash will go, for anything, from a Smart to a Brooks, and I have listened, for the last time, to the one about "being practically a member of the family."

To be continued, we hope.

One afternoon a few weeks later, a stocky, somber-looking Negro, quite bald, and middle-aged, but obviously of great strength, appeared and asked me for a job. He looked clean, competent, and mature. He gave as references some neighbors for whom he had worked for years, and who recommended him in the highest possible terms. It seemed that he had had only intermittent jobs for several years, and was glad to get my poor ten dollars a week. I hired him, and he has been with us for a year and four months. I hope he will be working here for many years more. His name is Leo James. Hubert had known me as a boy, and it seemed right for him to call me by my given name, but this man calls us Mr. and Mrs. Smart, and we call him James. This seems to confuse the neighbors, but not James himself.

After his previous appearances in these pages, which I hope won't offend him if he ever reads them, he needs only a few additional remarks. After a year, we raised his wages a dollar a week. I don't think eleven dollars is enough for any man, at any job, but this is all we feel we can afford, right now, and there are some perquisites, including the use of a car, a gallon of whole or lightly skimmed milk a day, and miscellaneous vegetables and clothes. Every Christmas he gets five dollars and some little gift of use. When we were away in New York, and he stayed in the house and did the chores on Sunday, he got a dollar extra per week. He comes at seven-thirty, brings his lunch, leaves at five-thirty, and earns every penny and every perquisite he gets. As these pages have shown, he is good at farming, chores, housework, painting, upholstering, elementary carpentry, and almost any odd job. When we were away, there were no accidents, and the animals all did well. The dogs all hated the Brookses, which was a nuisance, but they are all attached to James. James does more and more shopping, and is intrusted with more and more money, and he never loses a penny. What is even more astonishing, he never has to borrow. I could go on like this for quite a while.

There is only one little matter in which James seems humanly feeble: we have to lock one sideboard door. Oddly enough, with a little whiskey or rum generally available, nothing ever disappears but blackberry cordial, or a little Grand Marnier, and once a little of my homemade wine, more precious to me than any spirits. I think now that it was not James who took this wine. We don't take any of this too seriously, because with a man working in the house, an unlocked sideboard is practically an invitation, and it is quite possible that he hasn't taken a drop. In fact, once or twice last winter, just before James left to go home in a sub-zero storm, I gave him a shot of rum. Even Captain Bligh did that much.

The cloud that seemed to hang over James in those first few months has practically disappeared, I am glad to say. I used to wonder about it, but I never lost any sleep over it, because the Brookses had just left, and because I remembered the remark of a sound man and a sound printer named Hal Marchbanks: "If I wondered much about the private lives of my employees, I'd have shot myself long ago."

Wages, hours, conditions, and "recognition" are another matter. As I say, I don't think James gets enough, but with our present uncertainty, I don't want to pay any more. That is, I'd rather do his work myself than pay more. The fact that he is getting slightly more, perhaps, than other farm hands seems to me somewhat irrelevant; besides, he is worth more than most. His hours are long indeed, but the work is relatively light, and he gets into town often during the day. When the weather is bad, we try to find work indoors. I'd be glad if there were a good farm wage-workers' union, which would have to be organized on an industrial basis, since there are few specialized "crafts."

However, I don't think there will ever be one. I look forward, rather, to the day when large farms will be owned and worked, as partners, by everyone working on them, without employees. I'd like to work on a farm like that myself. Whether it would be possible within a capitalistic society and state, I don't know.

Meanwhile, all I can do is hope that James finds his job a good one, and that we don't go broke.

Homage to an American farmer.

In discussing the farm at Bellbridge and its operation, and in certain other connections, I may or may not have revealed the fact that to me personally, quite aside from our economic relation, one of the most important men alive is a farmer named Sam Kincaid. I doubt whether he has, or will ever have, any idea that this is true, and it might not make much difference to him, in any case. There are so many years, so many facts, and so many experiences between us, as well as those that bind us together, that it is with considerable daring and presumption that I call him my friend. In so far as we are friends, it is I who am honored by that relation, and proud of it. It is hard for me to speak of this relation, because I fear that Mr. Kincaid might be embarrassed and baffled by my doing so, yet I must, because this book would be false and incomplete without a short sketch of the man himself, and without some poor expression of my homage. Here again I must be forgiven for introducing a little family and personal history.

Almost every visit of my immediate family to Oak Hill has included a trip to this farm at Bellbridge. My memories go back to a time when my grandfather was dead, and my grandmother, a very old lady, could no longer make the trip. My father, mother, sister, aunt, and I would drive up in a carriage. It was a big expedition, an adventure, for which we got up very early. On the way up the valley, my aunt would point out all the old farms, and tell us scraps of their history. My father was a conversational and absent-minded driver, and we were in almost constant danger from the new motorists dashing along at twenty miles an hour and scaring our horse, or from the railroad and threshing machines, which provided even greater terrors. Once at the farm, the horse was stabled, and the children were turned loose.

At the time I speak of, the tenant was an impressive old man with a white beard, named Mr. Albert Kincaid, who lived on his own adjoining farm with his active little wife, three grown sons, and two daughters. We were always given a gargantuan dinner, and after that, Mr. Kincaid Senior, my aunt, and my father started on a long walk of inspection. I usually tagged along. Once, they say, I fell in the creek with all the absurd clothes of a child then, was fished out by Mr. Sam Kincaid, and was put to bed, raging, until my clothes were dry. Late in the afternoon, bearing many gifts, and always something special for my grandmother, we started on the long, adventurous trip home.

One somber November Sunday, more than a year ago, Peggy and I and some friends went up to the farm on a picnic, and afterwards we took a long walk with Mr. Sam Kincaid's niece, who pointed out to us for the first time the lonely little house, whose boards have now turned silver, where her grandparents first began to keep house. They had been pioneers from Virginia, and there had been wildcats in the woods. As we wandered through those lonely upland fields, glens, and woods it felt like wandering back into the boyhood of Abraham Lincoln. I should not have been at all surprised if a tall frontiersman in a coonskin cap, with a long rifle, had appeared at the edge of the wood, stared at us, and quietly disappeared. We lost all track of time, got back after dark, and had to do the chores by lantern.

When the senior Kincaids died, our farm was taken over

by their oldest son, Sam, who lived at his old home with another bachelor brother, and was helped regularly by his sisters, who came out from town. The two brothers, the third, married brother who lives nearby, and a nephew band together to work on their farms and on ours. They are all thin, tough, brown, bent little men, with large noses and bushy black eyebrows hung low over dark little eyes. Of this rather formidable crew, who have, individually and collectively, the look of being hard to beat, Sam seems to have the most intelligence; he certainly has the greatest authority. They have all slowly grown more friendly, but they are still very shy, without any timidity. When we appeared, the other brothers used to say "How d'ye do?" politely, and then go on with their work, or disappear.

My grandmother had died also, my father, mother, and sister appeared less frequently, and the old era had definitely passed. When I was here, my aunt and I made the trip with friends, or in a hired car. The dinner had become a picnic at the falls, and the walks of inspection were made only by my aunt, Mr. Sam Kincaid, myself, and our dogs. My aunt always told me as much about the farm as I would listen to, but that was not as much as it might have been, because I was just a young literary man from the city, and I was always more interested in the creek and its birds, the moving clouds, the hills, the beech wood, my aunt, Mr. Kincaid, and the relation between them, than in fertilizers, the prices of corn, wheat, and hay, floods, and the prospects for the year. Even at that time, Mr. Kincaid was something of a hero to me, and the quiet amusement with which he regarded me and my own work was never unfriendly.

Despite what I have long thought were economic absurdities beneath it, the relation between my aunt and Mr. Kincaid was anything but absurd. As I remember that small, practical, alert, and well-informed maiden lady, in her early sixties, and the tough little old bachelor farmer, tramping about their fields together, I see the source of the superficial and maddening argument that "it isn't the system, but the people," and I marvel at the ability of good will and good sense to humanize and prolong obsolete economic techniques. The relation between these two was based on mutual respect and courtesy. It was strictly limited, but very deep. Each respected the character and competence of the other, and there was deep human sympathy between them, but aside from business, or rather, from production, the mutual ignorance was complete. There was a little gentle humor, but for the most part, the relation was extremely sober. I dare say that no man except her father and brother ever meant so much to my aunt, and yet Mr. Kincaid never even took a meal in this house, and I don't imagine this idea ever occurred to either. This strict limitation was intuitive, ingrained, and perhaps right, but it does seem to me a little sad.

When my aunt died, Mr. Kincaid did not receive a telegram from us, through a grotesque inadvertence that I shall always regret, but read of it in a newspaper. The day my mother and I arrived, he appeared at Oak Hill, uncomfortably well dressed, but very simple, matter of fact, gentle, and deeply moved. We asked him to be one of the pallbearers, and I hoped that pleased him. As we went in and out of the church, and to the grave, I felt that that bowed and somber little man, so strong, gentle, and remote, was there, really there, more than anyone else. At any rate, one of the few things that kept me sane, in that macabre and elaborate hocus-pocus, was the thought of those two walking in the fields and sunlight with their dogs.

A few days later, my sister and I drove to the farm, to see how things stood, and to pay for some paint that my aunt had had on her mind not a week before, when she and I had gone to the farm for the last time. It was a painful visit, but I shall always remember it for Mr. Kincaid's behavior, which was intuitive and exquisite. There we were, my sister and I, one more generation of landlords. We were twenty-eight and thirty, and he was, I should say, nearly sixty. We had his life work, and a large part of his livelihood, in our hands, and he really knew very little about us. Throughout our walk and talk he recognized instinctively both that we were so much younger than he, and that, most unhappily, we had to do business with him, and had to be taken seriously. His behavior has never fallen below that exalted level.

We renewed the arrangement with him, and took a long walk, discussing this and that. My heart felt like a cold stone in my body, and I don't suppose Mr. Kincaid felt any better. Fortunately my sister, ordinarily as reserved as my aunt, has a talent for sensitive gaiety, almost flippancy, when a situation is difficult. She joked with Mr. Kincaid in a way that I could never equal. I remember her asking him what he did at night. No Smart had ever asked a Kincaid such a question, and I guess he relished it. An almost merry light came into his deep-sunken eyes, and he said: "Well, I read the paper, and then I'm ready to go to bed. You see, I get up at four." On our walk, I slung my coat over my shoulder, and unwittingly dropped my wallet. Later he and I walked back, and by retracing our steps, he found it. "Well, I reckon that's a good thing," he said, smiling; "I reckon your ma would have given you a licking."

Since then, the relation has been much the same, superficially at least, as during the last thirty years. I go up oftener, because we can drive, and I like any reasonable excuse for talking with Mr. Kincaid. We go up to get a little corn, or to see the wheat threshed, or to see about some paint and fencing, or to consult about Johnson Grass, or to see the latest disaster wrought by drought or floods. I am acutely aware of my ignorance and uselessness, but I do whatever I can to be of help. Once, in acknowledging a check, I told Mr. Kincaid briefly and simply why I was damned if I could see what right I had to his checks. Except for very oblique and tentative little jokes, he has ignored this completely. He has always called me by my given name, and now, rather timidly, I do the same to him. He settles on the nail, and I try to do the same. I send him cigars at Christmas, and he sends us the most wonderful sausage and pork tenderloin when he butchers. I don't see why our partner never comes to see us, or why we are afraid to invite him up here for a meal, but then there are so many things that I don't understand, and that I'd rather handle intuitively than emotionally and rationally.

When I first appeared with Peggy, he was very pleasant, and very reserved. It made me feel odd to see him take off his hat, and be careful not to swear, and say "Yes, ma'am," as he did to my aunt. Luckily, Peggy has my sister's talent for ignoring his abysses of reserve, without invading them, and often, now, they are quite merry together. We almost always go to Bellbridge together, with all four dogs. When I go up without Peggy, Sam always asks about "the little woman," as he calls her.

I don't think I am a snob, even an inverted one, romancing about this "poor but honest working man," this "nature's nobleman," and so on. I have known enough working men to know that manual labor, poverty, and even honesty are no more likely to result in quality than are leisure, wealth, and family trees. But the fact is that whenever I introduce my friends to Mr. Kincaid, I wonder not what they will think of him, but what he will think of them. If there is anything false or cheap in them, it may appear, and I am always just a shade nervous about this rigorous test of myself.

This man's quality is not easy to define, or account for. It is not merely his fine instincts, honesty, competence, cleanliness, and long life of severe but not unhappy toil. It may have something to do with his deep reserve, his solitude, and his cunning with the soil and with all living things. Mr. Kincaid knows almost nothing about anything but his own kind of farming, and his life lacks completely the sweet and saving drugs of art and thought, as well as many simple comforts and pleasures. It may be that I find in this man an example of that moral fiber, very fine yet unbreakable, that I may lack myself, and that any good American civilization will require.

In January, after postponing it as long as we could, Peggy and I went up to look at the results of the latest flood. I don't have to repeat that tale of disaster, but I may remind the reader that about a third of the productivity and value of the farm had been destroyed, and that it was quite clear that in about one more little flood, the creek would cut off and destroy most of Mr. Kincaid's livelihood, and ours. We tramped about, rather glumly, and discussed various schemes, none very good. Mr. Kincaid saw it all, very clearly, and looked even more bowed, his face even more deeply furrowed. But he was going to do what he could with the holes, and farm what he had left. My contribution, relatively futile, was to try to get federal assistance, and to keep agitating for the Scioto-Sandusky Water Conservancy Project, and for the Ohio Valley Authority. Before we came away, we looked again at Mr. Kincaid's colt, Nellie, which he had rescued, with her mother, from her barn, and Mr. Kincaid gave Peggy some of his fresh sausage. "We'll do what we can, and you come up again soon," he said, and we drove away. He turned from us and from that desolation, and went about his chores.

Often we wonder about the future. Last June, Mr. Kincaid discovered that he had an internal abscess, and had to be taken up to Columbus, to a hospital, for a fortnight. The treatment was successful, and he looks better than he has for years, but the experience was a nightmare to him, and to everyone who knows him. Peggy and I called on him, in hospital, a couple of times, but there wasn't anything we could do for a man quite wretched because he had been outdoors all his life, and had always taken care of himself. When he was able to come home and watch the threshing, every man, woman, and child there was happy with relief. The notion of Mr. Kincaid's getting old simply makes me shudder.

The most likely possibility is that ancient Paint Creek, that sinister beauty, will one day take the shorter, easier way, and give the whole farm, once more, to the reeds, to the bass, to the herons, and to the grave and lovely meditations of the cottonwoods and sycamores.

Whatever happens, what little I am worth, in any way, is behind Mr. Kincaid, to the end.

An idea for some Galsworthy, but not for me.

I have mentioned the road up this hill as being as bloody as the sunken one at Waterloo. I am now forced to record those struggles, and their result for the time being.

When I came out here to live, I inherited a quarrel over this road that had been going on, intermittently, for almost a hundred years. George W. Dun had bought the land—we have the receipts—and had built the road, with a foundation of heavy rock. His descendants had kept it up, at considerable expense, and had always claimed exclusive ownership and use. Deeds had become confused, and records had been lost, and this claim was constantly disputed. No settlement had been effected by a couple of lawsuits indulged in, and lost, by my grandfather. My aunt was superlatively gentle, and neither possessive nor acquisitive, but she felt with some reason, in this road matter, that some of her neighbors were trying to take advantage of her solitude and sex, and she refused to yield an inch. She was therefore repeatedly inconvenienced and insulted, and the whole affair became, in her mind, almost an obsession.

During my second summer here, the whole matter was

opened once more by the plans of some newer neighbors to build houses along the road, and to use it. The question was raised decisively by the building of a culvert designed to permit a new, shorter road to join ours.

One hot summer day, Peggy came home from town with the news that the building of this culvert had begun, and that all the neighbors were assembled there, watching it. With a heavy heart, I took Jack and walked down to talk with them. I hate quarreling over property, I hate negotiation, and as you may have gathered, the whole idea of private ownership is repugnant to me. On the other hand, I did own property, and I represented both my living sister and her children and all our dead. I didn't propose to be cheated and imposed on, and I had not forgotten the pain that some of these people, notably the Murgatroyds, had caused my aunt.

As I approached, an unpleasant silence fell on the company. A role had been forced on me that I detested, but there was no escaping it, and I decided to play it as well as I could. I greeted them, and told them they might as well save their concrete, because I was going to see my lawyer, Joe Rowan, that day, and had no doubt of his ability to get an injunction stopping the work. We could not avoid going over the old arguments, briefly, I got a lot of nasty looks, and the assembly broke up. The work on the culvert was stopped immediately.

There followed a whole summer of painful negotiation, which resulted, finally, in the signing, by all parties concerned, of an agreement, recorded in the courthouse, that opened the road to all adjoining landowners, provided for its upkeep, removed gates and livestock forever, and imposed certain restrictions upon the houses that could be built on the adjoining property. The matter of actual ownership was not settled, or even affected, but agreement was reached on a number of more important, more practical matters, that would not have been settled by any decision of any court. Everyone gave up something claimed or hoped-for, and got something in exchange; everyone understands now at least where he stands; and everyone escaped litigation. My own lawyer's fee was small, but I had to spend twenty dollars on fencing, and that hurt.

At first, the human side of the negotiations interested me keenly. I noticed that as the summer wore on, with long and inconclusive conferences on the road itself, in lawyers' offices, and by telephone, the fight settled down to a struggle of wills between me and one neighbor, Murgatroyd. We were both determined to get everything we could outside of court, to force each other to the extreme limit, and as time went on, the points debated became mere pawns, of no importance in themselves. What I feared in Murgatroyd was an almost hysterical stubbornness that might well force us into court despite his lawyer. What Murgatroyd feared in me was memory and pride, armed with he did not know—fortunately for me—how much money. He is probably much richer than we are, and certainly lives more simply and safely. All the old wounds were opened up, and bled freely. I became haughty and sarcastic, and he became sullen and abusive. The only things that saved us were our common hatred of spending money, and the relative reasonableness of our lawyers, and of the other parties to the quarrel. But it interested me to notice that the emotions were not wholly useless. If either of us had been less bitter and stubborn, concessions might have been made that would have been regretted later. As it is, we explored, I think, most of the sources of trouble, and each of us is satisfied that he could not have got more, in or out of court. It also interested me to notice that the very age of the quarrel lent it a certain dignity, and that both Murgatroyd and I treated the newer people involved with a certain exclusiveness and condescension.

All this was interesting, yes, but only up to a point. It occurred to me more than once that the whole thing was like a story by John Galsworthy: quite as human, quite as well built, and, to my mind, quite as superficial and unreal. I kept thinking, here is something that would make a good story, but I couldn't think of it except as a neo-Galsworthian story, and that possibility left me quite cold. I had to recognize that possessiveness, pride, revenge, snobbishness, and all the rest were both human and important, but in the end, without fire from something else, they did not seem to me exciting. I kept looking, in vain, in all this vivid material, for even a bodiless abstraction, like Irene Forsyte, for something like "the element of beauty impinging on a possessive world," for something living or dying. It wasn't there. Refine, coarsen, distill, purify, build up, complicate, simplify, dramatize as best I could, the whole thing remained nothing but a dirty little road row, a grubby little quarrel between neighbors.

Our three new neighbors are a crack chemist at the paper mill, a hydrotherapist at the Veterans' Hospital, and a man in the County Engineer's office—all three married to women quite as distinctive as themselves. Literally, they are suburbanites, but actually they are not. I mean, something seems to have saved them from all that falseness, that bourgeois timidity and cheapness. It may be their jobs, which are not, you notice, buying and selling. It may be that when they came out to look at their new land, some essence of wildness stole up on them from the woods and blackberry patches, and saved them. At any rate, they are making their homes mostly the way they want them, regardless of fashions and other people's opinions, they wear overalls when they feel like it, they go down to the muddy creek for a swim, and they rent their extra land to farmers for cattle. Already some of them are wondering whether they can't keep a cow, and maybe a few sheep! Conventions are strong, but nature is definitely stronger. We have called on all of them, and they have all called on us. They let us collect their garbage for my hogs. One Sunday a couple of months ago, all six of them came walking up the road to tea, and Peggy and I, at least, had good fun. We did not, I confess, invite the Murgatroyds. We are on quite decent terms with them, and hope to remain so; there are a few scars that last, and ache a little, now and then, for life.

Thistledown on the wind.

We have three older neighbors who live farther away, but who for several reasons are closer to us than many people, in town and elsewhere, whom we see much more often. All of them have appeared already in these pages, but they cannot be mentioned merely in passing.

One of them is Ed Wagner, the market gardener and general farmer, who has worked harder, all his life, than almost any man I have ever known, who seems literally to be dogged by bad luck, and who still faces the next dawn, the next job, and the next disaster, with good cheer. Ed always has about ten irons in the fire, but any one of them is hotter than any of mine, and he always makes time to lend a hand to me or to any other neighbor in a pinch. But Ed not only works and is a good neighbor; he does something much more rare, he uses his head. While he was plowing my garden last year, fast and well, he and I kept up a running conversation about the electric company, the Farm Bureau, the new livestock auctions, taxes, local, state, and national politics, the rise in the price of horses, the defects of gin, the federal prison, and half a dozen other matters that interest us both. He had thought well, on a basis of factual knowledge, about all of them. Doubtless other farmers have thought as much, but they are often so close-mouthed that you'd never know it. Of course I have known Ed, off and on, most of my life, and it takes about ten years to lubricate most rural friendships with honest, fluent talk. If I knew exactly why Ed, his wife, and his four nice children aren't six times as well off, in every way, as they are, I might have a much more complete idea of "what this country needs," besides the good men it has. I get especially exercised about him because he has a joy in living that many farmers with more property and money seem to lack.

Another of my good neighbors is the Mr. Ralph Stone who has already been mentioned in several connections. Now Mr. Stone is a good deal like Mr. Kincaid in several ways, and arouses in me much the same excited and slightly baffled admiration. Besides the orchard to our south, he owns a very large farm in the valley. He has at least half a dozen men working for him all the time, and three or four occupied tenant houses on his place. Whenever he can possibly house and employ someone, he does so. He is careful with money to the point of seeming very tight, but I suspect him of constant secret generosities; above all, he is scrupulously honest. He married a nurse, and has three very nice children. He is on a township school board, is one of the pillars of the Farm Bureau, and goes, I think, to the Presbyterian Church.

When I was a boy, I always looked at this man with childish suspicion, because his uncle, many years before I was born, had shot my aunt's first Jack, an Irish setter, asserting that the dog had killed some of his sheep. My aunt had found the body at the foot of the hill, with no wool in the teeth. My aunt liked this younger Stone, and always said he was one of her best neighbors, but the image of that dead dog stayed in my mind for years.

Now ever since we have lived here, the entire family has been most friendly to both of us. Mr. Stone has sold me straw and hay, quickly, in a pinch, at market or less, he has let me hire men and machines, he has overseen the building of a line fence, keeping meticulous record of costs, except for omitting his own labor, and he has advised me carefully and thoughtfully whenever I have asked for advice. (As always, I think most of this rises from regard for my aunt; sometimes I think an inherited good name is as hard to hold on to, and as irrelevant, as inherited money.)

I shall always remember a bitter cold day in January when Mr. Stone drove into the yard, all dressed up, and in his very slow and careful way, asked me to join the Farm Bureau. I kept asking him into the house, but he said he didn't have time. Shivering to the bone, I seized the opportunity to raise every objection to the Farm Bureau I had ever heard of, or could possibly imagine, and got very good answers to all of them. I had long meant to join anyway, but I had never had a chance to argue with one of the moving spirits in it and in the Coöperative. I shall return to this whole question shortly. We also talked about the A.A.A. decision, about the Canadian Treaty, and (gingerly) about politics. Like many leading farmers in the county, Mr. Stone is, I should say, in the center or slightly to its right, politically, and unconsciously extremely Red in economics, sympathies, and general attitude. It will take many more disasters to their purses, or a Fascist drift in the cities and towns, or a war of fairly obvious uselessness, but I think that when the time comes, these lean brown graying men, or their sons, will be out with their shotguns, if need be, for production, use, and life, as opposed to property, profits, and death. This is only a guess, I admit, fathered by a hope. Well, that day we moved quickly on to the feeding of steers, and the care of sheep.

"Had any trouble with dogs, Allen?" he drawled.

"Not yet, Mr. Stone," I said. I find it impossible to say "'Ralph,'" in the country fashion.

"Well," he said, "you can't be too careful. When you find a strange dog anywhere near them, you get out your gun and shoot to kill."

He looked at me and smiled. I'd give a great deal to know exactly what he was thinking about, at that moment. That lovely setter had died forty years before, and I was warmed by that smile to my shivering marrow.

Another man whom I think of with Mr. Kincaid and Mr. Stone is Mr. Gabriel Oak, who went out with me that time to look at Mr. Finchley's sheep, and who, like Mr. Stone, has done me countless kindnesses. Mr. Oak is a dairyman and general farmer whom also I have known for years. There is another strong tie through the two Oak sons, who, when I was a little city boy on summer holidays, took me in tow, tried to teach me how to do a hundred things on a farm, laughed kindly at my failures, took me swimming with them, and generally made me feel I wasn't a quite hopeless weakling. One of them is an official at the prison now, and the other works at the paper mill, and we find it very hard to see enough of them. When I first knew them, their mother was dead, and Mr. Oak had not yet acquired his second wife, who is charming, and who leads everything good in rural society. Mr. Oak and I have shared my purebred ram, and Mrs. Oak has undertaken to help Peggy get started again this spring with better hens. Like all good farmers, these have to work themselves stiff, and we are tired at night, too, so we don't see as much of them as we'd like, but we know they understand.

I shan't soon forget a very few words I exchanged with Mr. Oak a few years ago, when I had just decided to give up my job and move out here.

"You're coming home to live?" he asked.

Whether I'm sentimental or not, I admit that those words, from that man, hit me squarely in the throat. I had never thought of it that way. It had not occurred to me that anyone else could think of it that way.

"Yes," I said.

"Good," he said.

Mr. Kincaid, Mr. Stone, Mr. Oak, and all the others like them, and their strong wives and children—they are not just my neighbors, and some day, I hope, my real friends. They are the America I love. One thing I definitely do not like about them is their quite unconscious faculty of making so very many people I know, who speak more of my language, look like children, dreamers, thistledown on the wind.

A few townsfolk, seen from a distance.

Of the twenty thousand people in Chillicothe, Peggy and I may know by name or sight nearly a thousand. This is a wild guess, based on my effort at one time to know all of the six hundred people in and about a school. One of the most obvious charms of living near a town of friendly people, near which members of one's family have lived for a long time, is the impossibility of going to town without seeing scores of pleasant acquaintances. Another is the ease with which one can meet, and keep on seeing, perhaps, anyone at all. We are far enough from cities to prevent the invasion or growth of any group of people whose principal social interests are elsewhere. I have been told that the snobbery is formidable, but I have never found it anything but slight, somewhat shamefaced, and amusing.

Despite all this, which I appreciate fully, I can say flatly that neither Peggy nor I can contemplate actually living and working in Chillicothe, or in any town like it, without a shudder. The reasons have been made clear, consciously and unconsciously, in thousands of novels, plays, movies, and magazines, and need no restatement here.

First I may remark that even discounting sentimental nostalgia for the past, there seems good reason to believe that the general quality of Chillicothe society has declined steadily from my grandparents' generation through my parents' to ours. The stories I have heard, showing incidentally the authentic culture, breadth, and refinement of my grandparents' contemporaries, would be hard to believe, were they not so well authenticated. In the generation of my father and aunt, there may not have been a marked decline in quality, but many of the most promising young men left town at the first opportunity, and many of the remaining women are now dead. The people our age and slightly older, whether they are the children of these others or not, seem to me in general to lack culture, breadth, refinement, individuality, and character. I can only hazard the guess that our ancestors built an economy, a society, and a culture which have now grown obsolete and shabby without our having the intelligence and courage to imagine, and begin creating, new ones. It may be suggestive that Peggy and I find the most exciting possibilities among those very few like Ed Brown (of the manhole story) who are consciously workers rather than owners, and who think more of the present and future than of the past.

But some of the survivors of the era preceding the present one of the D.A.R., the American Legion, and the Junior Chamber of Commerce are very dear to us. I think especially of our friend Miss Mary Yates Bell, who, more than she can imagine, has warmed and lightened our whole adventure. Miss Bell is a retired schoolmistress who lives alone in a nice old house, with a still nicer old garden, and who, with wide and deep interests, wit, detachment, and good will, quite without self-pity, and on very slender means, maintains a standard of inner and outer living, of civilization, to which Peggy and I can only aspire. She is our good companion, and she has done us a thousand little kindnesses, both practical and imaginative. In addition, she sees everything in long perspective. For instance, when we talk about our adventures in the Little Theater, she can tell us that as a fat infant my father once played Moses in the bull rushes! She and her people, now, with one brilliant exception, dead, long knew, loved, and forgave, all of my people, and these feelings were returned. These are little things, and I do not live in the past, but they can astonishingly enrich the present. I have wished that Miss Mary, with all her youthful spirit, had more faith and interest in the future, but as of Mr. Kincaid, that is really asking too much.

There is what might be called a group, composed, with their wives, of the executives in the paper mills, the owners of the larger independent businesses, and the professional men. Most of these have incomes ranging from comfortable to large, are rapidly working themselves to death, and spend their leisure playing bridge and golf with each other. We know many of these people, and they are very friendly to us, but our purposes and interests are so different that we naturally see little of them. It has occurred to me once or twice that if I could afford the time and money to run around with these people, I could learn much, and perhaps write some important and interesting short stories, but I doubt whether I could sustain a clinical interest in the forced and pathetically deluded shoots of a dying plant.

Yet in this group we have found a few exceptions or typical specimens that are really friends in that they meet us halfway. Notable among these are a paper mill executive and his wife whom we met (significantly enough) through Miss Mary Bell and through the Little Theater. Chauncey had been places, and seen and done things; he was, for instance, with Hoover in Belgium. Also, he is glad to argue with me endlessly and sportingly, about politics, economics, and sociology. In our thinking and talking, he suffers from a glut of facts, myopia, and a relatively large income, and I suffer from their opposites. When he has the time and energy, he is a grand comrade on the boards. His wife has a strong and charming individuality that suggests, in many ways, the triumphs, in character, of an earlier generation.

Among people more of our own age, there are several overlapping groups, based chiefly on friendships dating from school days, and not greatly altered, because these people do not seem to have grown or changed much since they got acquainted in high school. Peggy and I are glad that we are so clearly not wanted.

The fact is that we have depended, for social life, very largely on the Little Theater. This is not surprising, of course. Sympathy based on hard work towards a common goal can, while it lasts, evaporate astonishing differences between people. (Peggy has just found this proved again in working with the Red Cross for flood refugees.) And when the work is as intimate, and the purposes are as revealing, as in the attic ward for patients of "footlight fever," the social effects, however evanescent, are miraculous. No one has any real identity unless he puts on his own unique show, and it must be that amateur theatricals are of a real help to people in this, the most important of their most private affairs. Otherwise they wouldn't work themselves stiff with fatigue, and pay the high price in organization work and in petty rows. We have seen from fifteen to twenty people enslaved by the opening of a curtain on an illusion, and they are a rum lot. We have seen as many stories and novels being composed, in life, before our very eyes.

There is that little shoe clerk, the son of a prison guard, who applies real talents to the creation of the rôles of criminals and underdogs. There is that engineer, Ed Brown, who burned up the boards in his portrayal of a hungry and harrowed cab driver. There is that old newspaper editor who spares neither words nor feelings in his insistence on acting that is *acting*. There is that son and heir of a department store owner who spends his days with rugs, toilet articles, buttons, and piece goods, but has the very soul of a trouper, and is never really alive until the stage is set, the house is full, and the curtain opens on a play in which he can "set his teeth in" another life. There was that keen, emotional, handsome, and very lonely young Jewess who sent it across the footlights at 100,000 volts, and burned us all up in the process. There is our friend, the giant Sam Gladstone, who earns his money by making tombstones but earns his living by doing the work in half a dozen organizations. It was Sam, assisted by our friend Gerald Rowan, who begged, borrowed, appropriated, bought second-hand with his own purse, and made with his own hands, almost our entire stage equipment. He is a born craftsman, but it may go deeper than that: a stage is certainly a better place than a cemetery. Then there is Sam's wife, Betsy, who used to be a schoolmistress, and who has taste and knowledge very much needed by our outfit. The same thing can be said for Gerald Rowan. He happens to be his father's invaluable stooge at the local home financing company, but he is a scholar at heart, and keeps the wires open to the world theater.

Sam, Betsy, and Gerald are our particular side-kicks. They are among the few people whom we can't frighten, baffle, or annoy seriously. They sit on our lawn, drink our milk, and make endless good fun of us and of our animals. Without them, we'd be in something of a hole.

Another of our most highly valued resources is the Rowan family entire: father, mother, daughter, and two sons, distinct individuals, yet one of those few families who make me wonder whether there isn't something in this family idea, after all. We impose on them endlessly, and quite without shame, because we can talk with them, and laugh with them, for hours on end. We don't fool them a minute, and they like us just the same. It has no basis that I can put my finger on; it is merely one of those currents of warm feeling that are important without being quite explicable.

More than a party.

Except for house guests who can "take it," we do almost no entertaining. In the winter we have little room, and the house isn't safe for people used to steam heat. Also, entertaining is usually beyond our means. Also, our friends here are too heterogeneous for large parties, and if we entertained them all properly, in twos and threes, we'd have to have two or three parties a week, throughout the open season. At that time we usually have house guests who have fled to us from calls and parties. Finally, as Peggy might add, I am damnably unsociable. I like to talk with people, and to share with them whatever we have at the moment, but entertaining, as an art and as a duty, does not interest me.

Still, understanding all this, and lending all hands, our few side-kicks come out from time to time, to our delight. And two or three times, as many as twenty and thirty people have appeared at once, piloted and fully cared for, by Joe Rowan. At first, we felt a little inundated, but soon it seemed just right, and very pleasant, to have the old house, yard, and garden swarming with people who seemed to be enjoying themselves. The first time, Peggy's first cocker, Jerry, went around lapping up cocktails that people left on the stone steps. He was none the worse for it.

The party that I remember especially was on our first Christmas Day. We had three old friends of my aunt, with whom she had often spent Christmas, out to dinner. We were proud that they were friends of ours too. We fetched them out from town, and just after they got here, while we were having mild cocktails in my workshop, a real blizzard blew up. The sky blackened, the air was filled with driven, fog-like snow, and soon all the windows were opaque. I made sure that my sheep were under cover, and then we sat down to dinner.

The table was very pretty, with old glass and silver, and Peggy's decorations. We gave the ladies silly little things, like a bowl of goldfish, or a ten-cent fountain pen, and they gave us chutney, veal loaf, and a handwritten and illuminated card, by Miss Mary Bell, that was a "Litany of the Beasts." We had a goose with all the fixings, and a plum pudding brought in flaming.

I imagine Peggy missed her large family, and the ladies must have thought of other times and older friends, but everyone seemed quite merry. For reasons I don't quite understand, I am abnormally insensitive to Christmas, Easter, and so on, and Peggy has suffered from this defect of mine, but that day I felt that there was in the air, somewhere, a blessing.

The blizzard grew stronger, and since we didn't know how we were going to get the car back up the hill, when Peggy drove our guests into town, two of them, with foolhardy gallantry, insisted on riding in the rumble seat, to make one trip only. They survived it, and Peggy just managed to get herself and the car home, but we had to decline, at the last moment, an invitation to an evening party at the Rector's. Just before setting, the red sun appeared, illuminating a weird, bright waste of snow and huddled animals.

It was a strange day, remote from the lives we had lived, from the people we were. It was so dream-like as to be uncanny, yet it was a very happy dream. We felt a little like Leslie Howard, in *Berkeley Square*, at that moment when he finds himself, dazed and dreaming, yet alert and excited, actually alive in the eighteenth century.

Visitors from the old country.

During the six months that I lived here alone, I had nine house guests; the next year, Peggy and I had twenty-nine; the following year, we had twenty-two. That is, we have had sixty visits from about forty-five or fifty different people.

When you consider that we live 780 miles from our families, and from all our old friends, in a place that has no obvious attractions for the tourist or vacationer, that has no éclat whatever, and that is on the route to nowhere, except, oddly enough, Mexico, you will see how extraordinarily lucky we have been. Because we go to very little extra trouble for guests, neither of us has found them wearing, except possibly in extremely hot weather. On the contrary, they have all brought much more, in every way, than they have taken away. How much they were moved by simple curiosity, and how good times they have had, on the whole, we shall have no means of knowing until we see how many return, how often. They have all given us extremely good times; they have been good guests, good companions, and bearers of rich freight from our old world, and from the past and future.

Doubtless the number and frequency of our house guests have had a part in keeping us to some extent aliens in this place, but if this is so, we do not regret it. Our old friends mean much to us, and so does what might be called our citizenship in the world. If, as we don't have to, we should have to give up most of the past, and become strictly limited or provincial in our personal interests, in order to become real natives here, we should prefer to remain aliens. We dare to suspect that the people here can benefit almost as much from what they can get from us as we can benefit from what we can get from them. They have helped us to put down roots; perhaps we can help them to break into flowers more native and lovely than ours.

I seem to have wandered oddly off the subject. To go back, the only way I can give an impression of this stream of guests is by recalling at random a few images and memories.

My first guest was the fifteen-year-old son of my old Boss

at school. The poor devil had sweet corn, lima beans, and tomatoes almost every meal. He read many detective stories, and we went to movie thrillers at night, and by the time he went to bed, a slamming shutter, or a swallow in his chimney, made him jump out of his skin.

Lewis Nixon, a faculty crony, has dropped in five times on the way to or from Mexico or the West. He drives at terrible speed in an open Ford, wears a beard, usually brings along some boys or another schoolmaster, and is quite at home with all our animals and discomforts. Once he arrived in a dense fog at four in the morning, with a couple of his lads. We all ate sandwiches and drank beer and tequila, and talked Mexico, until dawn.

The other schoolmaster who often comes with Lewis is an English gentleman who sniffs our strange Western air with delicacy and zest—and who, like Lewis, warms my vitals.

Bob Allston, another schoolmaster, and his wife, despite their rather glittering and successful lives in Manhattan, have visited us twice, once in the dead of summer, and once in the dead of winter. Bob does things like making a snowplow, or helping Dr. Ames and me with an operation, and Arabella has a passion for pushing a lawn mower. We all like to sit on the lawn and settle the affairs of the world. They have brought their cairn, Craigie, and now they talk of their bairn, Sambo.

Another pair of friends, of our own age and sort, are Clive, the newspaper man, union leader,* and ex-cowboy, and his unique wife, who both have that ultimate sophistication of simplicity and candor. It seems to come from vitality, brains, and direct experience of divers realities, without accumulation of irrelevant mental baggage.

I think also of a visit to me in hospital of a writer who is a

*Since I wrote this, Clive has inherited a large warehouse business. He moved in a few weeks from leading one strike to facing another as an employer. Where could this happen but in the U.S.A.?

sort of scientific Dr. Johnson. His specialty is the humane criticism of machinery. His wife is a Peruvian, and they were on the way to Mexico.

At the same time, when I couldn't enjoy them, Peggy had a couple of New England ladies who adapted themselves perfectly to what was to them the Great West. One of them made some virile and satisfying paintings, and offered to scour, card, and spin a fleece or two!

Of course children have given us some of our most unexpected and memorable moments as host and hostess. I think I have already mentioned Peggy's niece, and my nephew and niece. Since I mentioned the nephew in a sort of George Washington and hatchet episode, I'd like also to recall a ride home from Bellbridge, one lovely summer evening, when this nephew George rode with Peggy and me, and improvised the most delightful little songs, all the way home. The world is full of Peter Pans, and their female counterparts, and I don't like them, but I wish that growing up didn't involve the separation of art from life, and of imagination from thinking. Their distinction, yes, but not their separation. I don't really think it has to.

On the general subject of entertaining relations, and relations-in-law, much of the time I think my good luck—in these varied and fascinating people, who are so wise and competent, who are such good companions and real friends—is most exceptional, and I feel appropriate warmth and gratitude. Yet I must confess that every now and then I wish that James Thurber would write and illustrate a little book called *The Octopus in the Bedroom; Or, Families and What to Do about Them*. I suppose the final answer is the attainment, somehow, by hook or crook, through hell or high water, of mutual respect and courtesy.

In fact, I suppose that may be the final answer to most human relations.

Neighborly dynamite.

I have told how Mr. Stone got me to join the Farm Bureau.

Three months later, we got a form letter, from the Bureau, inviting us to a meeting at the armory in town. Before the meeting, there was going to be a supper for the members. Someone had contributed a quarter of beef, someone else was going to roast it, and someone else was going to provide beans and coffee. The only ticket of admission was one pie per family. Peggy was recovering from the loss of some teeth, and—the symbolism did not wholly please me—neither of us eats pie, but we were damned if we were going to miss that meeting. The stores were about to close, so we got Gerald Rowan to buy a pie for us, Peggy got up and dressed, and we went.

The beef was excellent. At first the only person we knew was Mr. Stone, and he made places for us, at the long table, beside himself. But soon people began to introduce themselves, and we did not feel lonely and peculiar. On the contrary, I felt that I was getting into my own crowd for the first time. I have never been a joiner, and these social feelings always surprise me. I soon discovered, a little to my surprise and relief, that Peggy was falling for it even harder than I was. The average age of those present was about sixty. They were mostly lean, brown older men with gray or white hair, keen eyes, and tough hands. There were some competent-looking women, but one got the impression that they had come only to keep their husbands company.

After supper, everyone went upstairs where the acoustics were foul. The Chairman was completely unintelligible. A young boy and girl, who had been sent off to a Farm Bureau camp by the Board of Directors, to learn about Coöperation—everyone said it with a capital—made dull verbal reports, and thanks.

Then the speaker of the evening was introduced. He was the lobbyist—this word was never used—of the Farm Bureau in Columbus. He was a keen little man about forty, and a very effective public speaker. He spread papers all over the table, and went to it, vigorously, for about an hour and a half. His remarks had no coherence, but he knew the facts about everything he mentioned, and he stuck to subjects of immediate and concrete interest to everyone present: the social security legislation then being debated in the State Legislature; the kind of legislators, in general, that we had elected, and what to do about them at the next election; public ownership vs. Coöperative ownership of electric power plants and lines; the situation, at that time, of the Farm Bureau Coöperative Insurance Company; the Farm Bureau fertilizer business; the attacks on the Coöperative movement; the frauds of advertising; the exemption of farm goods from the state sales tax; and so on. The man was a realist, and obviously an effective lobbyist, yet his whole attitude towards the Coöperatives was nothing less than religious. He quoted the Bible at the drop of a hat, and seemed to make some of his most palpable hits in this way. One got the idea that Jesus Christ was the First Coöperator, and would have fought the utilities. I didn't know anything about that, but I could see that the others made the same connection between religion and this economic movement. True or false, it seemed to me a psychological connection of the first importance. I had heard about radical-religious-realistic farm leaders, and I had long admired a much greater one of the same kind, Secretary Wallace, but I had never seen one in the flesh. I felt I had got my five dollars' and one pie's worth in one evening.

I discovered later that the Farm Bureau and Coöperative hostility to "politics" is profound. I went to another similar meeting towards the end of the Presidential campaign, and politics were not mentioned, even in private conversation. (I did see one man with a Roosevelt button, and dared to ask him why so many farmers seemed to be anti-Administration, even when it had saved their farms for them. He whispered that everyone hates anyone who has loaned him money or helped save his life.) I have never even dared to point out to anyone but Mr. Stone that the Farm Bureau was in politics up to its neck, and that there was no essential difference between our lobby and another lobby I, at least, hated—that of the veterans. When I did so, to Mr. Stone, he said merely: "I know, but do you know any better way to get things done? And if we don't do something for ourselves, no one else will." As I have said, I think any attempt to hook the Farm Bureau definitely to any party will fail, and for the present, rightly. Only when the legislatures and the courts swing into action in defense of profits, and against Coöperation, will the Coöperatives see the light on this point.

I must explain that the Farm Bureau is not itself a Coöperative, in this county or anywhere else. It is primarily a parent of Coöperatives, and a propagandist organization. One can belong either to a Coöperative, or several Coöperatives, or to the Farm Bureau only, or to both. Last year, I spent a little more than a hundred dollars at the Farm Bureau Coöperative elevator, as a registered customer. A 5 per cent dividend for that year was declared, which gave me one share of stock. At the end of this year, I shall get whatever dividend is declared in cash, according to my purchases. I have not yet attended a Coöperative meeting, so am unable, to my great regret, to report a meeting of an organization that interests me almost more than the parent Farm Bureau.

But I have learned a few things of interest. Our local Coöperative elevator, handling grain, fertilizer, feed, fence, farm machinery, and the like, did last year a gross business of about a quarter of a million dollars. Coöperative prices are always market, but have several times proved their effect on the private elevators here. In my experience, the quality of the goods equals or exceeds that of private dealers. A closer alliance between Coöperative and Farm Bureau memberships has not been deemed advisable; they are considered, rightly I think, to have different functions. We are likely to have a Coöperative filling station this spring, and we can already buy motor oil. The greatest extensions, lately, have been in electrification and in insurance. Coöperative electric lines are not allowed to parallel private lines. It is not possible advantageously to change life insurance from a mutual policy, like mine, to a Coöperative policy. Fire insurance rates are lower, but the Coöperative insurance agent was not interested in taking a risk on the isolated buildings at Bellbridge, now protected, in part, by policies with private companies. The insurance on this house is another matter, but I had paid up the private policy for some years. In other parts of the state, many other forms of Coöperative purchasing have been undertaken successfully. There are plenty of good books and magazines on this subject in general.

A few weeks ago, I was delighted when Mrs. Ralph Stone called up to say that she and her husband wanted Peggy and me to come down to their house to a "covered dish supper," the next night, to be followed by a meeting of an "Advisory Council" of the Farm Bureau. We were in the last rehearsals of a play at the Little Theater, but Peggy offered to direct that next rehearsal, and I accepted the invitation. At the large and pleasant Stone farmhouse there were about a dozen or fifteen men and their wives. The men included Mr. Oak—who was one of the founders of our Coöperative—the manager and insurance agent from the Coöperative, and substantial farmers whom I had met elsewhere. I felt rather light in weight, and something of a tyro and interloper, but not because of anything said or suggested by these men and their wives, who were all most friendly. A schoolmate of my father's, now a large farmer and sheep man, was cordial as always, and though I am rather shy, I should feel at home and safe anywhere, if Mr. and Mrs. Stone and Mr. and Mrs. Oak were there. The supper, served from the dining room, and eaten on card tables, was really superb, and incidentally, I risked emphasizing my urban lightness, so to speak, by being the only man who sat down at a small table with women. After supper there were good songs by a high-school quartet, including the oldest Stone boy, and his sister, in a very pretty costume, entertained the company with some tap dances. Then the meeting began. Mrs. Oak was the secretary, and it quickly appeared that the women in this group were quite as active and important as the men. Three or four men read intelligent and interesting answers to questions about agricultural Coöperation that had been assigned to them at the preceding meeting, and a general but coherent discussion followed, in which the women took part, and in which the general attitude was sympathetic of course to Coöperation, but intelligently critical of it. If people like these are threshing these questions out, in this way, all over the country, something very important is happening. This time, the religious connection was not made, and the radical implications were once more ignored, not deliberately, I felt, but rather because they had not been grasped, and were still somewhat irrelevant. I need hardly say that for me this was one of the happiest and most interesting social evenings I have had in the last three years. The law of diminishing returns may be effective here, but I was glad and proud when Peggy and I were made regular members of the group, which meets monthly.*

A few days ago I had a long and interesting conversation

*Later bulletin: We have attended two other supper meetings of this group, and they have been just as interesting and pleasant. At one, one of the questions was: "Are you satisfied with the present economic system?" This question was approached gingerly, but the unanimous answer seemed to be: "No! But we don't know anything certainly better, except Coöperation, and we will not tolerate dictatorship in any form." American farmers may prove to be the bulwark of democracy, but I doubt whether they will be fooled into thinking that this implies the profit system.

in a barber shop with the manager of the Coöperative. He is an extremely keen fellow, and I wish I had several pages for this conversation. I may say that he said he left his father's farm because he was worked too hard as a young boy, that he thought most farm boys left home for the same reason, and that he thought farm boys would continue to do the same until rural electrification and other new developments could brighten farm homes and lighten farm labor. Incidentally he, like half a dozen farmers, mentioned a picture of my lambs that had appeared in the *Gazette,* and once again I had to explain that it had been taken and used only because the farm editor was a friend of mine and found my sheep the nearest. But all these men have been most friendly about this incident, and have acted (with a saving light in the eye) as though I were a real sheep man.

If the Coöperators here are not aware of the profoundly radical nature of their business, the private business men are, and I shall not be surprised by more violent and articulate opposition. A few months ago, two young insurance men called on us, with their wives, and I seized the chance to get their response to the Farm Bureau insurance. It was restrained, but violent. These young men seemed convinced that their own futures lay with "the big, old established companies." They are good men, and I hope they will learn before they are too old. One of the best things about the Coöperative movement, whatever its outcome, is the fact that it will force reconsideration of the profit motive, the profit system, and all the rest, by townsfolk who would otherwise dismiss any criticism of them as "idealistic and subversive Communism, utterly un-American." I can be dismissed as a crack-pot, but Mr. Ralph Stone cannot.

The future of the Coöperative movement will be extremely interesting. From what little I have read, I cannot quite understand the relations between the Coöperatives, private business, and the state, in the Scandinavian countries, and I cannot quite imagine the inevitably different future of the movement in this country. I can only guess that a time will come when the profit-makers and their deluded stooges will show their teeth in the legislatures and the courts, and when the inevitable movement of the Coöperatives back towards the sources of supply will encounter a last-ditch fight. At those critical moments, or at some others now impossible to foresee, the Coöporators will do well to have read Marx or his translators, as well as the Bible. At that moment they might examine the possibilities of alliances with the trade unions.

Meanwhile, the whole thing saves me money and gives me a periodic emotional jag that I wholly trust. My father and both my grandfathers were liberals and reformers, but as far as I can see, except for their personal characters and influences, they got precisely nowhere. I do not forget that they defended civil liberties at cost to themselves and their families, and that one had his printing house sacked, and was burned in effigy, for his pains. But when I sit there in those meetings, with those tough old men whose work is my work, and whose enemies are my enemies, I feel that we are going one step further, beyond ballots and editorials, into dollars and cents. I feel that we are there to have a hand in our own destinies, as producers, consumers, and men of this Republic, and that we are out to win.

"But neither are men brothers by speech—"

The isolation, here, of this radical attitude would be hard to exaggerate. Ross County has, I think, about forty thousand people, and I have not met more than a dozen whom one can imagine even to have considered it seriously. I know only one man, the Ed Brown of the manhole anecdote, and of the Little Theater, who can be said to share it. (The last time I saw Ed, his whole structure of faith and belief had been shaken, slightly, by the Trotsky trials, and much more seriously, by the fact that one of his subordinates, a "hill-billy," was revealed to have lived in incest with three young daughters, and given them all syphilis. Ed's socialism is or was based on a somewhat mystical belief in "the dignity of man," which is a much less firm basis than my more selfish one.) I remember especially my excitement about, and the general indifference to, the annihilation of the Viennese socialists. About three years ago an old gentleman, for many years friend and banker to my family, flattered me by remonstrating with me for my radical talk, which, he said, coming from a man in my position, could have a bad effect on the working class. This is so rare that I treasure it. Some of my remarks seem to have stuck like barbs, and festered, under certain skins, but in general, among my acquaintances of all classes, my political opinions are considered an amiable or unattractive eccentricity or weakness, bordering on the insane.

At the Little Theater I directed and produced a few episodes from *Waiting for Lefty,* by Clifford Odets, and this adventure had a few curious results, but since my interest in the play was primarily dramatic, I shall report it in another connection.

Now, philosophically and emotionally I am convinced that truth is what works, and that until the facts step in to prove or disprove beyond question, the efficacy and hence the truth of an idea is considerably a matter of agreement. A lunatic is simply someone who finds himself in a minority of one. I have the kind of mind that is not completely reassured by books, by letters from friends seven hundred miles away, or even by what I consider "the sensible and true avouch of mine own eyes." In politics, though in little else, I need, not too far away, what might be called brothers in belief. After a while this isolation began to get me.

About this time I heard from friends in New York of an organization, for communication and propaganda, whose ideas formed an American Marxism not much unlike mine. I am not naturally a joiner, but I began to feel the urge. The publications of this organization had an alarmingly academic tone, but I decided not to be too fussy, paid my dues, and got my card as a member. There was no "Unit" of the organization within a hundred miles, so finally I grew restless, and arranged an appointment with a Mr. Wilkins, the national organizer, who was going to be in Columbus, fifty miles away, one Sunday a year ago last November. I was particularly eager to get more light on the organization's farm policy. I had looked in vain (and am still looking) for a discussion, both radical and realistic, of American agriculture. I was also curious about the national organizer of a radical outfit.

Wilkins was a roly-poly, genial, ingenuous-looking little man of about forty—and not half so ingenuous as he looked. One might have taken him for an assistant professor of economics in a state university, discharged for his opinions. We met at a hotel that was, ironically, full of National Guard officers on a spree, so we departed quickly and found a quiet place for lunch.

My disappointment was immense. To my astonishment and chagrin, Wilkins knew even less about American agriculture than I did. He had traveled widely, and kept his eyes open, but he didn't know even the commonest facts about such things as the A.A.A., the Conservation Act, the sphere and influence of the County Agents of the Department of Agriculture, the Grange, the Farm Bureau, and the Coöperative movement. He admitted his ignorance with the most engaging candor, but this didn't help me any, and I didn't even find out the things I wanted to know about the organization. Our talk quickly developed into a cross-questioning of me by Wilkins, whose questions varied from the most general to the most personal. In the end, he made it clear that he would very much like to come down and meet some of my farmer friends and our County Agent, but he was keen enough not to press me on the point, and although he was fluent in the radical jargon (such as "C.P." for Communist Party, "Trotskyism," etc.), he was also keen enough not to call me Comrade. He doubtless pigeonholed me as an unreliable liberal.

Well, I pondered all this all the way home through the November rain, and in the barn, and out in the mud heaving corn fodder, and back in the dining room with Peggy, Sam, and Betsy, listening to some exquisite Wolf songs over the radio. My first response was simply "To hell with it." I had no inclination to use my personal relations with Messrs. Kincaid, Stone, Wagner, Oak, and others, including the County Agent, to ask them to educate a prominent, candidly ignorant, and lily-handed radical. Yet I soon saw that it wasn't as simple as that. I felt that if Wilkins was ignorant, so were they, and if he were a bit closer to them in work and general attitude, they could learn as much from him as he could from them. I thought of the farmers of Italy, Germany, and Russia. I knew that the effects of my decision would be submicroscopic, in any case, but I felt that because of the implications, my responsibility was great. I worked myself up into a lather. Finally I decided that mere ideas, however fertile and useful, could pass effectively only from one worker to another. If I, who was so much closer to these men than Wilkins, didn't dare argue with them, what chance would Wilkins have? I wrote to him, as tactfully as I could, that his education in American agriculture would have to begin elsewhere. He took it with excellent grace, and every few months, for a while, we exchanged letters about this and that.

However, last autumn the organization got out a circular outlining an organized attack, through Congressmen, on the Supreme Court. I am no defender of the Supreme Court, but I had read enough history to decide that the historical and theoretical background of the attack was flimsy indeed.* The whole thing had a juvenile tone, and sounded like a popgun. I welcomed the demand for unanimous support, and resigned from the organization. I wish it well, but I think it is up the wrong street. These ideas can pass only between brothers, and

> . . . neither are men brothers by speech—by saying so:
> Brotherhood here in the strange world is the rich and
> Rarest giving of life and the most valued:
> Not to be had for a word or a week's wishing.†

*I thought much the same, later, of a more formidable attack from another source.
†A. MacLeish: "Speech to those who say Comrade."

VIII. Fun

Organized illusions.

If the life here pictured has any one inclusive purpose, it is growth, and that seems to me to demand integration. Anything growing seems to have to be of one piece and texture, to function as a whole, without unrelated parts to harden or fester. Successes and failures in this direction will have been noted by the reader in what has already been reported.

The common separation of pleasure from the rest of life, and the title of this chapter both force me to emphasize here the fact that the activities I am about to group together as "Fun" are merely developments of other interests and activities, and are therefore not to be considered in and by themselves. One of the most effective reasons I came out here to live was that I wanted to have more fun in my work, and more work in my fun. And not only that: I wanted to have all the realms of activity suggested by these chapter titles involve all the others. Incidentally, I wanted to lessen the power of money, a medium or technique with which I have never been very familiar or successful.

After our work, in all its ramifications, we have probably got the most fun out of amateur theatricals. Since Little Theaters are all fairly similar, and have been much written about, I may be forgiven if I report only our personal experience in ours. In doing so, I do not mean to imply for a moment that our own work has been the best or most important. On the contrary, I imagine that our experiments and vagaries have contributed not a little to the disorganization and debility of the Little Theater of Chillicothe, Ohio, Inc. We have fairly good rooms in the attic of an office building, we have fairly good equipment, and we have twenty thousand people to draw from and to appeal to. Our troubles have several familiar causes. The most important is that we have no individual able and willing to give all his or her spare time and energy to pulling it together and making it go. Others are faddishness, juvenility, small-town sensitiveness and traffic in personalities, the over-organization of the town, snobbery, the lack of general interest in the theater as such, and so on. In talent and knowledge we are deficient, of course, but no more so than almost any Little Theater in a town of this size. If someone with these, with tact, with a passionate devotion to the theater as such, and with a will of steel, came to Chillicothe and took over the Little Theater, within a few years the whole theater world would know about our attic. I imagine this can be said of a thousand American towns.

When I first showed interest, I was drafted as an actor, and played a fat-headed Englishman in a three-act modern comedy. We gave a couple of public performances in a movie theater whose boards have been trod by Joe Jefferson, Booth, Maude Adams, and other great figures of the past. The production made up in verve what it lacked in finish, and we all had enormous fun. The play was generally considered sophisticated and dirty, and I was the only person to protest that it was neither. I was delighted by the experience of acting, but I was much more interested in watching the problems of direction, setting, and production, and their solutions, and in watching what got across the footlights and how. A year or so later, I played a backwoods-man in a one-act farce in the Workshop, and later in a church, but my chief interest has been, and probably will remain, in learning how to write for the theater. Just as others are given chances to extend themselves into parts they cannot play in ordinary life, so I am given chances to extend myself as an am-ateur playwright. Peggy is interested chiefly in direction and in settings. I persuaded her to walk on the stage as a maid in one of my little plays; she played it properly as a lecherous sloven, but with such care and accuracy, and with such a wag of the bottom, that in a minute and a half she brought down the house and nearly wrecked the show. She has refused to act again.

My first play was a little dialogue called, clinically enough, An Episode of Growth. The drama was entirely psychological, the lines were cryptic, wordy, and heavily weighed down with allusions obscure to the audience. The setting was good, but as a director, I knew almost nothing. The players put it on almost by themselves, with skill and enthusiasm. The first performance, at the Workshop, was a complete flop. That strange little play floated out across the footlights and settled, like an ice pack, on every forehead in the room. The audience was completely baffled, stunned, depressed. When I read H. G. Wells's account of Henry James's theatrical fiasco, I felt I understood it well. Later we put this play on before a women's club, complete with salute to flag, ballots, and violin solo. The ladies got a collec-tive headache, and the Little Theater got five dollars.

Later I wrote and put on another little piece, called Little Men. It was a satire, both on small-town life and on impracti-cal immigrant farmers such as we are taken to be. It involved a sister, three brothers, the sister's boy friend, and a dog. There were more action, more obvious humor, and more general intel-ligibility. We all had excellent fun, and the play was a modest success, but I perceived more clearly the dangers, in small-town amateur theatricals, of politics, profanity, and sex. The whole thing seemed, to me and to the players, quite sweet and innocent, but it left a Disagreeable Impression. This time it was the dog, and Peggy's backside, that had the most obvious effects.

My last effort, so far, has been a little philosophical fantasy called *Poor Richard*, from which politics, profanity, and sex are absent. Two naïve but educated old maids, living in the country with a moronic and demented middle-aged brother, involuntarily entertain an escaping convict, who allies himself with the phantasies of the moron, and talks himself out of the situation and the house, with some ordinary clothes. The whole thing is a variation on Pilate's ancient but still interesting and sardonic question. This piece was the most successful of the three, and my experience, as a whole, seems to suggest that amateur playwrights, in small towns, would do well to avoid realism, to concentrate on action, and merely to imply whatever interests them most.

However, there is no general rule, and my most exciting experience has been in directing some episodes from *Waiting for Lefty*, by Clifford Odets. (The playwright kindly allowed us to use these free.) Although the framework and the shouting from the house were most effective, settings and costumes were subordinate, and everything depended on the acting. I had a cast of people who were workers and knew it, and who could act. They sent those naked, hot, and desperate lines burning across the footlights. The house was small, and composed mostly of women. I think every hair present was stood on end, but aside from sympathy and alarm, I could read little on the faces.

Later I was accused, falsely, of putting the play on without authorization, and the president of the Little Theater said that if he'd been there, and known what was going on, he'd have resigned. It appeared later that he had been taken for a ride, on the subject, by his fellow American Legionnaires, and by his fellow members of an organization known as Americaneers. The latter interested me enormously, and I have been trying ever since, without success, to find out more about it. Such members as I have located are shamefaced and secretive, but it seems that it is or was an outfit of young bucks, run by our leading capitalists, for the purpose of illegally escorting labor agitators out of town. Remembering, among other things, the attack on my mother's father as a Copperhead, I was eager to have these outfits make some move against me, but nothing happened. But the episode from *Lefty* may not be over yet. I keep thinking that of all pants that are fun to kid off, toy uniform pants must be the most fun. Some day I hope to do a little one-act musical, about the Legion and the others, with a rousing chorus, "The Americaneers have hairy ears," and so on, not too clean and polite.

This spring, Peggy directed a one-act crime play, and then we both worked together to produce and direct a modern three-act comedy, in the Workshop, for members only. Peggy is much better than I am on stage movement, on settings, and on handling the women and in sympathizing with their passionate interest in their costumes. I am better than Peggy on the creation of characters, stage speech, and the handling of the men. We were extremely lucky in a hard-working, intelligent, talented, and congenial cast. They did most of the work of all kinds themselves, and so I can say that the final result was the best amateur performance in Chillicothe for many a year. We were also lucky in finding a new German lady of talent and charm to help design the setting and to paint the backdrop. In the last few days of rehearsals and stage-setting, her brother arrived from Germany, fresh from persecution by the Nazis,* and unable to speak more than three words of English. But he was a

*Just as some of our best people came from Germany in 1848, and from Ireland later.

trained decorator, and went happily and skillfully to work. We were less lucky in our audience. We gave two performances, and only about sixty members, in all, appeared and paid their dues. Perhaps they were discouraged by my malodorous reputation. In New York, the play was criticized as being a little sweet, a little Puritanical. Here, the audiences found it difficult to laugh,* and there have been mutterings about its having been "blasphemous." Sometimes we'd like to tell Chillicothe, Ohio, to go blow, but we have too much fun with it for that; it might well tell us to do the same.

I should like to register, here, a brief protest against the royalty policies of agents. Interventions by gallants like Odets are most infrequent, and agents seem to have a mental picture of Little Theaters as wallowing in money. We, for instance, haven't had more than a hundred dollars at any one time for years, and fifty-dollar royalties would wipe us out. Agents are also obscure and inconsistent in their policies, and boorish in their manners. The result is that fairly decent people are forced into a furtiveness that they find most distasteful. Largely for this reason, Peggy and I have decided that any other plays we direct will be by playwrights who (with their agents) are definitely dead, and whose plays are definitely alive.

Sometimes our Little Theater itself seems quite dead, but it never is. It may fold up, finally, some day, and we may not always have James to take over our work when we have to rehearse. Then, like most of the people in the world, we shall have to create all our illusions, make all our theater, in our own heads. But unlike most people, we shall be able to remember the opening of the curtain on our own first (and last) nights, on our own plays. And that, as the old lady remarked when she opened her boiled egg, is already something.

*Does movie-going destroy the participation of audiences?

"Adults, 39¢."

One of the alleged, and in part actual, defects of country life is its remoteness from the arts. It is true that Chillicothe is no longer visited by good companies on tour, that music has an uphill fight, that painting, the graphic arts,* sculpture, and architecture have here almost no public at all, and that even books are hard to get. Chillicothe has long had pretensions to culture, whatever that is, but I have to admit that they are now chiefly pretensions. However, what I always say is, every man his own artist, and in the country and in small towns, he is luckily forced into the entertainment of himself. Besides, we and most of the American people are in—generally before Broadway —on the rise of a great art, the movies. Since we have lived here, Peggy and I have become rabid movie fans, and are proud of it. I generally find that most of my brilliant ideas on this subject have been anticipated by Gilbert Seldes, or John Mosher, or Katharine Best, or somebody, but I have a few I can't restrain.

What interests me most in the movies is the creation and effect of phantasy, its slow refinement, deepening, and transformation into fantasy, and its gradual retreat before a more important art that is slowly being forced to rely on, and become, itself. In no respect do scholarship, education, and criticism seem to me more dead and futile than in their general scornful neglect of what may well be considered, in a century or two, the most important art of our time. While thousands of feeble candidates for degrees meekly spend years in digging up useless facts about completely dead businesses, business men, sciences, scientists, arts, and artists, here are a business, a science, and an art that as much as any are molding our characters and our culture, making us what we are. Without having studied the mat-

---

*Although Dard Hunter is probably our most distinguished citizen.

ter at all carefully or systematically, any good movie fan could ask a hundred questions about the movies that it would be very well for both the movies and the public to have studied by competent minds and answered as well as may be.

For instance, I'd like to read a short history of the movies in terms of its stars, written not to expose their private lives, but to make a guess at their functions, in relation to social history, as instruments of fantasy. To what extents, relatively, at what times, and why, have they become hits? Could the intricate and subtle effects of the movies on life outside the movie-houses be at all effectively tracked down? What about the much more rapid progress of the technicians of all kinds, compared with the writers and directors? Why aren't there more writer-directors? What about these teams of writers, and these "conferences"? Is there any chance of movie-houses becoming more specialized, as they already have in cities, and for westerns? How cheaply can a good movie be made? That is, what are the chances for experiment out from under the present movie entrepreneurs? What are the chances for movie libraries and repertory houses? And so on, indefinitely. I am aware that there are occasional magazine articles on these subjects, and that sometimes these are good, but I wish there were more and better books in this field, and that there were a movie magazine comparable in quality to *Theatre Arts Monthly*.

Everything considered, it seems to me that the achievements to date are astonishing, and that the promise for the future is very great. If we live long enough, we may see pictures beginning, rather than ending, in a clinch; we may see secretaries who don't marry their bosses; we may see failures, heroic and otherwise; we may see triumphs that do not involve money, or sex, or fame; we may see good artists that stay poor and beat their wives; we may see the causes, rather than the detection, of crime; we may see animals in the cycle of their lives, without hunters, or photographers, or their damned women; we may see men at work and not in love; we may see children being natural and interesting, instead of theatrical and dull; we may see fantasies closer to Lewis Carroll and Voltaire; we may actually see ourselves, not as we'd like to be, when we get tired and hopeless, not even as others see us, but as we are seen by men who have attained, for brief moments, what I can only call the vision of God.

Meanwhile, and this is all I really have to say, you don't have to live in New York in order to see, first-hand, one of the most exciting and important things in the world today.

A footnote to Veblen.

In Thorstein Veblen's *Theory of the Leisure Class*, there is, as I remember it, an attack on competitive shows of purebred animals, on the ground that they are examples of "conspicuous waste." In this, as in other matters, such as the attack on the printing of William Morris, Veblen reveals a certain inhumanity and a certain ignorance that weaken a brilliant book. The genetic utility of animal shows is obvious, and the very human pleasures they offer cannot be neglected by any sociologist. These shows should be criticized when "show points" become distinct from utility and beauty, and when cheap exhibitionism, at the expense of the animals, supplants sporting competition.

So far, we have only entered one dog in one show, and since nearly everyone has shown, at some time or another, a dog, or a cat, or a bull, or a sheep, or a horse, or a vase of flowers, or a can of tomatoes, or a postage stamp, or a green cheese, or a figure in a swimming suit, this experience will be familiar.

A few weeks before the show, which was at Dayton, eighty miles or so away, Peter developed a sore lip that promised to rule him out. He responded to treatment, but until the morning of the show, we didn't know whether to show him or not. Besides medical treatment, he got several baths and much combing, brushing, and clipping. He didn't like this, but he liked the attention, and his personality flowered as a result of the whole experience. He had been shown many times as a puppy, and was getting back into his own. We went to a wedding and reception the night before, and had to leave home at six-thirty in the morning.

The show was at a fair grounds, and beautifully managed. There were some five hundred dogs of all breeds, and none were benched. The people were mostly veterinarian-looking men and women, and very nice, with a few people of the tough and sporty type, not so nice. The breeders and exhibitors looked very much like their own dogs; at least, a chow owner was just as unpleasant as a chow, and all the women with toy dogs were just as disgusting as their pets. (Peggy and I kept feeling our ears, to see if they were growing.) The judging of cockers was unanimously approved. The cocker people were polite but reserved, and I thought I felt a general undercurrent of suspicion and jealousy. If we knew more dog people, shows would probably be pleasanter, but perhaps we'd be more conscious of that undercurrent. There was no humor at all. Dog-breeders seem to be very conscious of the importance of being earnest. The atmosphere might have been that of a surgical lecture, or a courtroom, or a church. When the court rose for lunch, everyone stood up with a sigh, and I fully expected someone to ask us all to unite in a word of prayer.

When Peter's number and first class were called, Peggy turned white, and my heart was pounding foolishly. She had been nervous about showing him, but he remembered perfectly everything necessary, and did himself proud. (Mr. Veblen, it is very nice to have a handsome wife showing a handsome dog.) There were four other dogs in the class, shown by men. After long delay, and tremendous suspense, another dog got the blue ribbon, and Peter the red. In the next class, Peter was unopposed, and so Peggy got a blue ribbon anyway.

We were both exhausted, and came away. We had learned a good deal, and talked dogs all the way home. When we got home, we hung the ribbons, temporarily, on the mantelpiece, and felt pretty good, thank you. All the other dogs were jealous. Peter licked each one Hello, and then went out and had himself a good roll in the leaves and dirt.

We plan to show Peter a few more times, to see how far he can go, and for the fun of it, but unless we go into the dog business, that will be all.

Assorted nostalgias.

I mean our longings to sail boats, play squash racquets, sing, dance, and ride horses. These longings—generally mild enough —are nostalgic either for our individual pasts or for the past of the race.

Peggy and her brother have sailed boats off Cape Cod nearly every summer all their lives. I know he's a good sailor, and I fancy she is; at any rate, it's in her blood. When we were married, a New Bedford paper mentioned the event chiefly because, as they put it, "the bride is a descendant of nine generations of men in the whale fishery." So the bride brought beyond the Alleghenies a mahogany sea chest, some nice little seascapes in oil, and some memories and hungers that she doesn't mention often, but that I fancy will be there for many years, and may ache a little, from time to time, in the right weather.

Though I have never had the remotest chance of becoming either, I have often felt like saying, with Hazlitt: "I am so sick of this trade of authorship, that I have a much greater ambition to be the best racket-player, than the best prose-writer of the age." Whenever, in New York, things began to get me down, all I had to do was to play squash racquets for a half hour, to wipe everything else more completely from my mind than I could in any other activity, bar none, to be completely refreshed in mind as well as body, and to start again with zest. Cutting firewood and plenty of other things do that for me now, but not quite so completely. The one and only time when I have been homesick for New York (as opposed to individuals there) was on seeing in a magazine a good photograph of two men playing squash racquets.

Though I have never been able to sing worth two cents, even to my own satisfaction, and even in a bath, I have some very happy memories associated with singing, and I consider this deficiency one of the minor curses of my life. However, for people of my sort, a farm rivals a bathtub. One never has that illusion of resonance, but one has privacy, and can bellow one's head off at work. James sings right well, like most colored people, but he is very shy about it. I am still rather shy about "singing" around the house, but in the fields and especially in the barn, my improvisations make up in gusto and shame-lessness what they lack in music. I can approximate a few sea chanteys, and an old Yorkshire song that I can only spell pho-netically, "Ilkla mor ba dat," but my masterpiece and favorite, especially while milking, is an endless piece very vaguely sug-gestive of a Gregorian chant. My barnyard Latin is both fluent and meaningless, and I attain what suffices to suggest to me the pure, sexless wonder and sorrow of that amazing music. It is deeply satisfying, especially on a somber evening of autumn or winter, with the milk squirting rhythmically into the pail, the mud and manure sucking one's feet, a wet tail smacking one across the face, and a cold drizzle permeating the old building, one's clothes, and one's very marrow.

"*Pater noster qui es in coelis, sanctificetur nomen tuum.* . . . So-o! Put your foot back! . . . *Agnus dei, qui tollis peccata mundi, miserere nobis!* . . . *Hagii Ichthyos quelque chose!* . . . *Arma virumque cano, Senatus Populusque Romanus, amo, amas, amat, amamus, amatis, Aa-mant!* . . . Stand still! . . . *Gloria in excelsis Deo!* . . . There you are; all over . . . *Ite, missa est.* . . *Ite, missa est.*

I hope that in writing about the movies I expressed or implied no scorn for the phantastic and vicarious pleasures of our fellow movie fans. Because by watching Fred Astaire and Ginger Rogers (whom I can't put in her partner's class), Peggy and I do the only dancing we can do. Peggy can appreciate and criticize the technical brilliance. All I can do is pay our way in, sit down in the dark beside Peggy, a dull and awkward clod, and for two hours be, vicariously, a witty young man, expressing with ease and subtlety a prodigious *élan*, gaiety, *joie de vivre*. It lightens my hours for days. . . . We have heard of square dancing, back-country, but we have never had a chance to try it.

Riding can be, of course, one of the major pleasures of country life. My aunt kept a couple of horses, and gave me a saddle and bridle. I didn't mind riding to town on errands, but I always preferred long jaunts over the back roads in the country. I learned what little I did learn on a horse that had only a walk and a hard fast trot. My legs were fairly short and weak for that, but I learned things I couldn't have otherwise. My aunt taught me a good deal about the care of a horse, but no one taught me anything about riding. The other horse, acquired later, had a nice canter. I didn't mind riding alone, in fact, I preferred it, but I used now and then to try to get people in town to ride with me. No one was interested, then. Now riding is all the rage, and we can't afford to keep horses.

Sometimes when some animal wakes me at night, I lie

awake for a little while, thinking of the times when, as a boy, I'd wake up at night to hear one of the horses snorting, beneath my window, get up, go to the window, and watch them moving in the moonlight. And I remember the beauty of those horses running, when I tried to catch them beneath great trees that now too are gone. And I remember my aunt's smiling, taking a rope and an ear of corn, and walking right up to them. And I think of those long rides alone, when the pain and confusion of adolescence slowly and surely disappeared, and the world appeared once more in its wonder. And I think of Mr. Kincaid, and his colt, Nellie. And I remember a line from *Oedipus at Colonus*: "The might of horses, the might of young horses, the might of the sea."

Of course one of the real delights of country life is an occasional expedition to a city. Often the anticipation is better than the fact, but no matter. Columbus is our nearest city, but except for a remarkable collection of modern paintings at the museum, the veterinary clinic at the State University, and one bar, it leaves me fairly cold. We went to Cleveland once, when Peggy had to go to the Clinic, so the circumstances were not the most favorable. Besides, while we were there, the city was invaded by half a million of those dear people, the American Legionnaires and their wives, and I had to cling very firmly to the idea that these were not representative of the American people, yes. However, we were most graciously entertained by some people who had not seen me for nineteen years, and the museum had a magnificent loan exhibition. In fact, I have a tentative idea that individuals and museums are the only things that redeem these inland cities, including Chicago. We have had several short and pleasant expeditions to Cincinnati, including the end (for me) of a Girl Scout convention and the one I mentioned in relation to the price of wool. On the latter occasion we spent twenty-four hours eating, drinking, bathing, dancing, going to the movies, and visiting the museum, from which we were thrown out because we had the presumptuous idea that a rule might be broken briefly in the interest of the public, i.e., ourselves. We stole a peek at an exhibition being prepared, and the Director, with a rudeness and arrogance I have not seen equaled, even in an aesthetic functionary, escorted us to the door.

Even in New York, somewhat to my surprise, our greatest pleasures came from the same things: friends, pictures, theater, food, drink, and comfort. Of course the pleasure of seeing most of one's old cronies, after two years, was enormous, and it was augmented, in the most miraculous fashion, by the arrival, at just that one time, of a couple of them from Europe. Peggy has never been a New Yorker, but I thought I might regain my quondam sense, in walking the streets, of being an understanding part of that glittering and somber spectacle. I never did, partly, perhaps, because we were usually exhausted by so much steel, concrete, noise, and rather hectic society, and by some difficult business and personal relations. Aside from being at home with families and friends, the only places where I felt at home were the Gracie Mansion, the Hayden Planetarium, the Harvard Club library, and a Cézanne exhibition. We reached the melancholy conclusion that we could not live happily in that city, and that we'd rather go on relief, or starve, at home.

R.F.D. 3.

But expeditions and visits like these cost money that we may or may not have again, and for the most part our knowledge of the Great World will have to come through our tin mail box on the turnpike. I have had what may be called two commencements. In 1926 I became an A.B., a classmate, by virtue of his LL. D., of Andrew Mellon, who controls our gouging electric company, and one of seventy thousand Harvard men. In 1934 I became an R.F.D., a neighbor of the Kincaids, Stones, Oaks, Wagners, et al., and one of the thirty million men, women, and children who try to feed themselves and America. (And I have been accused, for my radical talk, of biting the hand that feeds me.)

We write and we receive a relatively large number of letters, and most of those that we receive seem to us very good. There are certainly more celebrated people than those who write to me, and there are doubtless those who write letters of greater elegance and subtlety, but the letters that my friends write to me are almost terribly alive with gusto, curiosity, anger, amusement, good will, wit, and *joie de vivre*. When I hear anyone say that the art of letter-writing is extinct, I smile. Letters may or may not be saved as they once were, but that doesn't seem to me of the first importance, and besides, any quarter of a century that produces, for instance, *The Letters of D. H. Lawrence*, is doing well both by itself and by later times.

We don't happen to save any, even of the best, even from those closest to us, or best known in the world. One reason is that in the attic of this house there have been trunks and basketfuls of old letters. My aunt destroyed a great many, and Peggy and I—never without reading them first—have destroyed many more. The job of reading them was not wholly unpleasant, and we saved a few letters of the early nineteenth century that seemed to catch some of the life of the time, in this country and in Scotland. On one of her visits, my mother tied these into neat little packets, with red tape, and labeled them. But from a mass of old letters there rises a sense of the insignificance of the little affairs that seem so important, of the transience and spiritual waste of property, of the power of disease and death, and of the futility and pathos of the human adventure. These letters were not more trivial, or materialistic, or sad, than most, but they contained so few of the sensations, ideas, experiences of the moment, so little of the life that alone justified all the hope, worry, scheming, disappointment, and petty triumph. They did not seem to do justice to all the lives they annotated and recorded. As we dumped them into a grate, and watched the flames blacken them, and then turn them into cool gray ashes, it was like hearing a thousand weary and happy sighs rising from every room in the house, upon the attainment, at long last, of utter extinction and peace.

Thank you, thank you, they murmured, leaving us, and I wondered how much of our own lives we were distilling into something worth putting into letters, something so useless, hopeless, and fearless, so immediate, and so unquestionable, that it would never make any difference whether it ever got into letters, whether or not it ever became nothing but soft gray ashes in a grate.

At any rate, my friends and I are out to live, now. We are trying to make and do things that please us, now. Our descendants can take care of themselves. If they want to read our old letters, they can, if there are any left to read. If they want to save our old junk they can, if there is any of it left. If they prefer to forget us entirely, they can, and there will be no sighing in the rooms where we have lived.

How to be a freak.

In several ways, these last sections raise the whole rather amusing question of being an intellectual in the country, in our time. Urban intellectuals see so much of each other, and so little of other kinds of people, that they are apt, I imagine, to underestimate their isolation, and perhaps to misunderstand their significance and function, as intellectuals. I shouldn't dare to define for them their function, but I can assure them that they are in an extremely small and isolated minority, and I can contribute my own small experience towards the making of definitions and tactics.

In the first place, I should like to clear away a few possible misconceptions.

One of our guests from New York was evidently surprised and relieved to discover that I had not become very self-consciously proletarian, and rabidly nonintellectual. It appeared that he had encountered a number of the sort of youthful, intellectual, and religious radicals that I so heartily dislike. I can only say that although—after individuals, who will always come first with me—I much prefer my working part to my owning part, and am very definitely for workers as opposed to owners, I know very well that I can work better with my brain than with my hands, and see absolutely no reason for trying to throw away my most useful tool. I can't save myself, or be of any use to anyone else, without using it to the limit, and with what little pride it may justify. I am almost as proud of being one of those seventy thousand Harvard* men as of being one of thirty million farmers.

Another guest from the city raised the same question in another form. He is a composer, and a highly intellectual artist, who, except for the most marvelous manual skill with instruments, can't do anything at all with his hands or body. He calls me an artist, and he evidently feared that, as such, I had gone over to what he calls "the tough boys, like Hemingway," or at least to "the gut boys, like Lawrence." I hope I don't have to say that although I admire men who can do things, and revel in my own slightly ridiculous attempts at action, and that although I admire men of intuitive and prophetic power, and try

*Cf. An Exchange of Letters, by Thomas Mann.

constantly to open myself to the "otherness" all around me, I know very well that ultimately both action and intuition are dangerous and silly without the best knowledge, the best sheer reasoning, and the most active common sense available. Absent-minded professors, chorus girls, and farmers, are more attractive to me than intellectual snobs, straight or inverted.

Now the isolation. It would be hard to exaggerate. I have already suggested the cultural poverty of the place, as evidenced in the mass and in institutions. I may point it up by saying that among the individuals we know, here, only very few may be called educated—using that word in the rather limited sense that one can discuss ideas with them, or refer to a book, or a picture, or a scientific hypothesis, or a piece of music, without terminating the conversation. This is all relative, of course. Among most of my city and faculty friends, my "culture" is almost pathetic, while here I am considered a howling intellectual and aesthete. Yet there is without doubt a kind of glorious freemasonry among those who are interested in ideas, sensations, books, and works of art, and who have begun to appreciate and define their own ignorance. There is that whole group of references which, as Aldous Huxley or someone has pointed out, are like a body of family jokes and skeletons, and afford much the same pleasure. In all this, then, we are almost alone, and feel our isolation more keenly than I expected we should.

The alternative, I suppose, would be a college town or an art colony, and both, I think, ask too much in other ways for what they give. Both are isolated from what I call reality, and are soaked in pettiness. Give me my farmers, old maids, and "footlight fever" patients, any day.

Besides, I suspect that in some ways this isolation is a very good thing. It is annoying, and even dangerous, in that it encourages mental indolence and indifference, and permits a dulling of the wit and the growth of a certain dreamy, romantic egoism. However, it forces the intellectual life to maintain and justify itself, and it forces the growth of personal relations that have other and perhaps equally important bases. Farming is immeasurably more sympathetic to thought than is business, for example, and it is more stimulating and vitalizing to it than is teaching in a school. And if my intellectual past now separates me from my neighbors, it may be that in time, as I learn to think in terms of this life, my intellectual activity will draw me to them. Even now, once in a blue moon, I have a feeling that a man of action is almost as fascinated by what is going on in my head as I am by what he can do with his body—and I don't mean in any false or snobbish way. Degrees, past or future books, and little areas of recondite knowledge mean absolutely nothing here, and I am glad of it.

Now then, under these conditions, what sorts of thinking are feasible and effective? What sorts of thinking *go*? The rest of this book will answer this question better than anything I can say consciously. I'd like, however, to point out a few possibilities that I have neglected, to my own cost.

First would come simple watching. As I have said, farming is so hazardous and absorbing that it is hard to keep one's eyes open to nature for herself. I'm afraid that after three years I could put down on one page a list of all the trees, shrubs, wild flowers, ferns, fungi, birds, fishes, snakes, wild mammals, and stars that I could name, find, and tell something about. I have no ambition to become an amateur naturalist of the school that memorizes the common and Latin names, throws in a little human psychology and a lot of sentimentality, and lets it go at that. However, I think this ignorance is rather shameful, and I have the greatest admiration for Henry Williamson and men of that kind (there are so few of them), who can destroy their own egos (or rather, let them flower) by projecting themselves into animals as remote as otters and salmon. Even more, I am shamed and irritated by my ignorance of the actual nature of the hundred-odd animals that are somewhat dependent on me for survival. I have here beside me a dozen or so pages that I have written on the non-human psychology of sheep, cattle, fowl, and dogs, and I have some ideas on the subject that please me, (e.g., on the individual-collective psychology of sheep, and on the social significance of dogs' excreta), but I want to know very much more before I talk in public. It may be that as time goes on I shall get a firmer grip on the farming, and so be able to look about. I sometimes think that one who had greater talents in this direction than mine, would do better to live in the country without doing any farming at all. He would be isolated, but good naturalists are rarer than humanists and radicals.

My next major regret as an intellectual is that I have learned so damned little, even about agriculture. I have done a fair of amount of talking about it here, because I had to, in order to show how it strikes an immigrant farmer three years out, but I know very well that if any of my neighbors has the generosity and curiosity to sit down at night, after hard work, and read this book, he will at least get a number of hearty laughs that I haven't planned for him. In fact, I think our only contribution to the countryside, to all these people who have lent us a hand, has been comic relief. We don't mind that at all, but I hope that in ten or twenty years we'll really know something about farming, both practical and scientific. If I sat down and wrote out all the questions I'd like to know the answers to, they'd fill many pages, and of course for every answer, four more questions pop up. The danger will come when, if ever, we can make a little money without learning more.

Another failure, another possibility, I have to record is in studying the people I do see. I tell myself that if I can remember anything, I can remember external details, and strong psychological hunches, about people. However, I have been so busy with my own little concerns that I have not yet got so deeply into any of these people as I'd like. I might be able to do neo-Mansfield stories about them, but that isn't what I want. I'd rather go through neo-Turgenev or neo-Chekhov into Smart. Incidentally, two writers I'd like to see at work in Ross County, Ohio, are Morley Callaghan and Thomas Bell. Peggy tells me that my almost willful ignorance of the great urban middle class is appalling, and I'm afraid she may be right.

Another failure, another possibility, is in daydreaming. I do not mean in sitting around and imagining what we'd do with a hundred thousand dollars, but in creating fairy tales whose basic ideas are most realistically true of Americans and of people anywhere, whose characters suggest the possibilities, and whose settings catch the beauty, of this scene. It always surprises me that aside from the Negro creations, we have dreamed so few figures like Johnny Appleseed, whom I happen to love, and Paul Bunyan, whom I happen to detest. At one time I found myself carrying on many conversations with a certain Mr. Abraham Hermes, who lives over on the Poke Hollow Road. I say conversations, but they were mostly endless yarns by Mr. Hermes. I haven't seen much of him lately; maybe I have been sitting too much at a typewriter; maybe he doesn't like my interest in economics and politics. But I don't think we are through with each other yet, by a good deal.

Another failure, another possibility, is in—but no, this is enough, perhaps, to suggest why living in the country practically demands the most ardent and happy intellectual life of which one is capable.

It may be noted that I have not even mentioned reading. It seems too obvious. Besides, the possibilities here need to be pointed out, not to the readers of this book, but rather to the sort of people who have been mentioned in it. Our general near-illiteracy seems to me quite as serious and dangerous as any of our economic, social, and political ailments. I think a free exchange of ideas, information, opinions, and experiences, and a fairly general and proper resort to books, for whatever they are worth, are absolutely necessary for the preservation of our freedom and democracy, such as they are, and for the attainment of any American civilization. I am not going on to a defense of the schools, and so on. What we need is not more education, but better education. We shan't get that until we get a socialist America, and we shan't get that until a few more good workmen (with collars or without) get a few more good cracks on the chin. They'll get them, all right, so I have an idea that some time after I'm dead, the Chillicothe Public Library will flourish.

I may say, to conclude, that in the last two years and a half I have found time to read, in addition to books and pamphlets on farming, about seventy-five fairly solid books, and about a dozen light ones. (Mr. Terence Holliday, bookseller extraordinary, of New York, gets entirely too much of my money.) These seventy-five have been mostly history, economics, sociology, and criticism, with a fair number of autobiographies and letters, a fair number of plays, and a handful of novels and poems. For knowledge, I have read entirely too little; for the keenest life in the moment, a little too much; for my farm and purse, entirely too much; and to be considered the county crack-pot, just about the right amount.

Of sensuality.

Perhaps I have made it clear by this time that a farm is a very poor place to earn a living in the ordinary meaning of that phrase, and a very rich and splendid place to earn a living in every other meaning of that phrase. I mean in the growing, so to speak, of ideas, sensations, intuitions, feelings, sympathies, and delight in action, of all those things that alone justify work and money. Here I want to make a few remarks about the senses, which seem to me quite as important, and quite as full of possibilities, as the mind.

I shall not indulge in a catalogue of sensual delights, because as a sensualist I am a city-bred tyro, and because I have always preferred actual to vicarious sensuality, and do not expect the reader to be any more disinterested than I am.

Neither shall I go into the morality of the matter. For that there are D. H. Lawrence, J. C. Powys, and others. That involves the proper integration of thought, work, emotions, sympathies, and so on, which, kept in their several places, can be great allies of the senses, and which, disintegrated and run wild, can dry the senses up. Obvious enemies of good sensuality are extremes of energy, ambition, and fatigue, anxiety, difficult personal relations, celibacy, and too much reading. Illness, on a farm, is very bad luck, but the enforced relaxation may even induce a revival and refinement of the senses. The whole moral question is complex and difficult, and all most of us can do is to think of sensuality as something precious, to be refined, and defended, and to keep as healthy and relaxed as we can. And here we have waiting for us, always, those mighty allies, the arts.

(In this time of death and rebirth, it seems to me that three things can save us: (1) a social reorganization to control land, tools, and money in the service of life; (2) a neo-religious and moral re-definition and revival, to release and apply the energies of the "men of good will" in the service of life; and (3) a re-definition and popular revival of thought and the arts, so that they shall not be strange and fearful, but shall intensify and refine everyday life. Not one of these can be complete or effective without the other two.)

I think it is curious, and may be worth noting in passing,

that literature, an art to which I have been attached in one way or another since I learned the alphabet, has not for many years been of great service to my senses. With the exceptions of Thomas Hardy and D. H. Lawrence, no writer has begun to do for me, in this way, what John Keats did for me when I was a boy—and I have not looked into Keats for many years.

On the other hand, painting, sculpture, the graphic arts, photography, the movies, and even architecture have become for me powerful drugs, that heighten rather than dim the beauty of reality. I really couldn't get along without them. I'd rather give up liquor, tobacco, and every acre of land, every stitch of clothing, every building, every last cent, and almost every book and animal.

We have a few pictures, originals and reproductions, that we love for themselves, and also for their effect on our response to everything around us. I have only to raise my eyes in this room, or to go downstairs, and in almost every city that I shall ever know, somewhere, hidden in the ugliness, and guarded by snobs, there will be marks of oil, and color, and ink, on canvas and paper and plaster, that will restore wonder to the earth. Every day I see shapes, colors, textures, and arrangements of animals, or trees, or grasses, or hills, or clouds, or merely dishes on a tray, enveloped in different kinds of light and atmospheres, that not only make me feel good, deep inside, but also, often enough, seem to have inarticulate, non-literary "meanings" of tremendous importance, merely by being what they are. Often enough, action fails, the human heart is cold, and the truth is both difficult and unimportant. Then it is that I have recourse to these things, these simple, useless, inhuman, insignificant things that demand no action, or sympathy, or understanding, but that simply exist, here in this room or out the back door, in their own wonder and beauty. (It is no surprise to me that so many of the best pictures have come from ordinary failure, loneliness, and bewilderment, from a very different and special success, warmth, and comprehension.) I wish that every man could be a real artist in something, and could share this attitude, which seems to me more important, more prudent, than, for instance, insurance. For this attitude we have to thank a lot of men, very strong in their own way, who in other ways were generally incompetent, lonely, and baffled, failures, spongers, ruffians, ingrates, and boors, without time or money for insurance, or even for bread.

I can imagine the scorn with which many intellectuals and aesthetes will read these words. "Such news! So art is wonderful! Really! . . ." I am not talking to these people. They don't matter. I am talking to good farmers, and to good workmen of all kinds. I am talking to those less lucky than I am. I am talking to the tired, the hungry, the worried, and the very busy. And I am saying Thank You, in my poor way, to all good artists, living and dead.

Sculpture could be good on a farm. I imagine it could make one more sensitive to the shapes of hills, to the beautiful bony and muscular structure of animals, and to the feeling, in the hands, of wood, steel, stone, concrete, bark, mud, dust, sand, loam, clay, leaves, stems, roots, horn, leather, wool, hide, hair, and skin, living and dead. But it is not possible, because it costs so damned much, and in a photograph a piece of sculpture loses most of its values.

Neither can I talk about music, first because I am densely ignorant of it, and second because it seems to speak directly to the nerves and spirit, almost without the medium of the senses. Of course it is a sensual delight, of the purest kind, yet it is very different, somehow. I have listened to a relatively large amount of it, but I can't see that it has made me more sensitive to sound, or increased my pleasure in other sounds. Yet when my friend the composer visited us, I could see how abnormally, almost pathologically, sensitive he was to all sounds. All I can say is that we have periods of playing the little phonograph for hours every day, and then neglect it for weeks together. (I detest the radio, and except for the symphonies and chamber music, can't listen to it for a moment without going berserk. This curious phobia works some hardship on Peggy and our guests.) Yet often, listening to a piece by Mozart, say, I feel that here, in this mere arrangement of sounds, all human life is caught, purged of all dross, and distilled into something of absolute purity, something dangerous and alien, yet made of our very souls, our only excuse, our only end.

Of course among the most obvious means to sensuality are the things one lives with. It is not only work outdoors, fun in bed, long walks across country, baths in the creek, meals, drinks, clothes, pictures, and so on, that we love, but also fine-grained wood rubbed clean and soft, well-woven and well-dyed textures of wool and linen and cotton, good silver, pewter, and brass, old and new, good china and glass, and fresh flowers in every room, whenever we have them. Here our tastes are abominably expensive, and we are abominably lucky. Sometimes I remember with shame that one of the most splendid sensualists on record built himself a house for $28.12½, lived in it for eight months for $33.87¼, and in very odd moments during that period earned $36.78. Still, I once lived for a month, in considerable sensual luxury, for about $10, and during that time earned about $12. When I consider my rich and poor friends in this light, I think it is a question of how accurately and lightly the money is spent, of how little one needs for a debauch, and of how conscious one is of one's luck, good or bad, and of its origins.

The ropes that bound Odysseus.

Of course the surest and cheapest way to multiply and intensify one's debauches is to pick up a pencil or a brush or a chisel and go to work, strictly as an amateur, and strictly for one's own private amusement. These last provisos are the ropes that bound Odysseus to the mast, when he heard the song of the Sirens.

I am not at all sure that this kind of "work" isn't as fruitful as any other. When I drew water all that summer, and carried it to my tomato plants, all the tomatoes rotted on the vines. I wish I had spent all that time and all that energy trying to make one good drawing of a tomato leaf. If I had, I should have no fewer cans on the shelves, and much more delight in my head, right now.

When I was recovering from my operation, I had to sit around a good deal. I got a drawing pad, and some crayons and pencils, and began to make pictures. Thanks chiefly to the paintings of Charles Sheeler, I had come to take great pleasure in the surfaces and shapes of ordinary buildings. Our Dower House had been finished a few weeks before, and the rear view of it, with the hen house and the privy, and some fences, posts, and gates, made a very pleasant pattern of light and shade. I drew a picture of it that was very bad indeed, but that fixed its beauty indelibly in my mind. I also made some sketches of sheep that could be identified as sheep, but that did not get what I especially wanted: the delicacy and strength of the legs. When Peggy had time, she joined me, and we made some caricatures of each other. She also made a nice little flower piece, and I made a pencil drawing of her, nude: the legs were too long, but the whole thing wasn't too revolting. We discovered that although neither of us had any real skill, Peggy had some feeling for color, and I for shape, line, and perspective. Too soon we were both too busy, we thought, for drawing pictures, but within a few months we were very glad we had made this little discovery.

The attic room used by our Little Theater was being redecorated by the members. The stage was rebuilt, with a proscenium arch covered with beaver board, and the entire room, after lengthy discussion, was painted, with water colors, a very pale green. Curtains were made and hung, and theatrical pictures were mounted and arranged. The whole effect was good, but something seemed to be lacking. Peggy and I decided that the arch was too bare, and needed decoration, both for its own sake and to balance the colored curtains at the other end of the room. Everyone agreed, but no one was willing to undertake the actual decoration. We had an architect who could have done it, but he was too busy. Sam Gladstone could have done it, but he was too modest. Peggy and I didn't know whether we could do it or not, but we were neither busy nor modest, and at the dead end of winter, we needed some distraction from our chills and farm problems. We gave the others a decent chance to object, and then got to work.

First I drew a sketch to scale, Peggy water-colored it, and Sam Gladstone made some much needed corrections. The temperature was well below zero, and we couldn't go to the West Indies, so we decided on tropical contents: palm trees, monkeys, flowers, and lush though stylized vegetation. For technique, we imitated Rousseau le Douanier, the decorations of a barroom in Columbus, and the pale colors of Puvis de Chavannes. The sketch was not detailed, and we made changes as we went along. The next day, by squaring it off, I drew the entire thing on the arch in white chalk. This was more tedious than difficult. The day after that, Peggy joined me, and we began the painting. We used poster paints, whitened and diluted, and fortunately Peggy had had experience with these. We found it hard to get the colors as pale as we wanted them, and at the same time, on that base, opaque. As it was, even after they had dried they were much darker and more intense than we had wanted them, and for a few moments, we were frightened. However, it soon appeared that because of the size of the arch and room, they were not too loud, and that if we had succeeded in making them paler, they would have looked washed-out. Our two major errors had canceled each other; when fools rush in, they sometimes have very good luck. In the sketch, the monkeys, bearing masks and thumb-nosing each other, had appeared pink, human, and vaguely obscene, so on the walls we tried to simianize their figures, and colored them violet. This helped, but one tail still springs from a buttock rather than the spine, and another monkey, Peggy insists, looks exactly like James Cagney, nude, painted violet. The total result was at first rather shy-making, but oddly enough, it has worn fairly well. As soon as anyone gets tired of it, he can wash it all off with a sponge and make another.

The mural is, we hope, impermanent, but the work itself was the keenest and most enduring fun. For three days we worked like laughing maniacs, felt like Michelangelo breaking his neck on the ceiling of the Sistine Chapel, got completely outside ourselves and all our winter fatigues and irritations, and came back to kitchen and barnyard feeling as though we had been in Martinique. The paints and brushes had cost us about five dollars, and after we had finished, this money was paid back to us by the Little Theater. There is more than one way to go South in the winter.

IX. Reflections

The only true education.

Like many others, this book offers ample evidence that country life can be conducive to prolonged if not always effective meditation on almost everything under the sun. A farmer's life is a busy and absorbing one, but most of his work cannot be hurried, and most of it is done in comparative solitude. As a result, he is apt to indulge in that rather amiable vice, soliloquy. This vice is not, I think, limited to immigrant farmers who are more literary and articulate than the others. I fancy I can see it in the eyes, and in the odd remarks, of Messrs. Kincaid, Oak, Stone, and others, and if I could see into their minds and get their meditations down on paper, I should have something much more curious and probably much more valuable than anything I have to offer here. But this is not possible, and in conclusion the reader is asked to look into four paths of thought down which I have wandered from time to time during the last three years. I am going to consider, very briefly, education, simplicity, acting, and death. On these formidable subjects I have nothing new, and certainly nothing final, to say; I merely want to suggest how they look to one dirty and absent-minded farmer, under the vast windy sky.

When I became a schoolmaster, I discovered the profound delight of teaching, even amidst luxury, in an institution designed to shelter rich people's sons from life, and to turn them into bogus English gentlemen.* I also discovered that the only effective teaching of the English language and literature that I did occurred when I and my pupils became, in part, in simulacrum, a workman and his apprentices at work on a useful job. When some job of reading or writing seemed—God knows how, in those circumstances—worth doing, and of some use to someone, and we worked on it together, then and then only was there a sound triangular relation between us and the job, then and then only was a passably good job done, then and then only did education, of both them and myself, occur.

This idea has come back to me again and again, here on the farm, so very far away from classrooms, from libraries, and from all the artificial charm of the sheltered academic life. I have thought repeatedly of what Thoreau had to say on this subject, and of the early experiences of Mark Twain, recorded in *Life on the Mississippi*. Mr. Bixby, Sam Clemens's pilot, one of the greatest teachers recorded in print, was not primarily concerned with his cub. His job was to get a boat up and down a dangerous river, and he had no time to worry about his cub's parents and complexes. He taught only when he had time, and then he went right to that job. He said: "When I say I'll learn a man the river, I mean it. And you can depend on it, I'll learn him or kill him." And the boy did not think of his master with the pitying condescension of a nice schoolboy towards his masters. After he had seen that master at work, in a pinch, he felt: "By the shadow of Death, but he's a lightning pilot!"

* This design is perhaps not conscious in anyone, and there have been enough exceptions and rebels, among boys, masters, and parents, to make the school a great one, despite the economic and social background, and their appropriate ideal. However, these of course remain powerful.

I am convinced that the only true education is that which takes place between master and apprentice engaged in a common task required by society. Schools and colleges, with all their remoteness and artificiality, are shortcuts, required and advisable at special times and in special cases only. I suspect that both Newman and Macaulay were wrong. Liberal knowledge seems to me neither the special privilege of parasites and escapists nor an obsolete perversion, but rather the flower of working knowledge, its purpose being to enlighten work, deepen play, relate work more closely to play, and civilize experience. Formal education, whether it knows it or not, is an instrument of society, and cannot be reformed alone. Therefore I cannot become passionately interested in experiments like those in the Universities of Chicago, Wisconsin, and Cincinnati, and in Antioch, Bennington, and Black Mountain Colleges.

I dream sometimes of a system of education for a new, a socialist America, and I have dared to sketch it in my mind, but since it is a bodiless dream, of interest only to pedagogues and radicals, I shall not outline it here.

Why, indeed, do I still think about these things? For two reasons. In the first place, even mediocre teaching, under luxurious circumstances, is a major experience, not easily forgotten and permanently discarded. Somehow, somewhere, I want to experience again that most exciting combination, or fusion, of working, learning, and teaching. In the second place, I see here, on this farm, an opportunity to do so. Peggy knows from experience what I mean, and both of us are working and learning all the time. Why can't we add teaching?

I certainly do not mean tutoring for examinations, like my tutoring in French a couple of summers ago. That experience, although pleasant and profitable, merely demonstrated vividly what I mean. The boy was spoiling his fun to get "credits" by studying a language that was an unadulterated chore, quite irrelevant to his life. I was neglecting my animals and place to get sure money by teaching a language that was to me a pure luxury.

We have in mind, rather, taking some summer a couple of boys and girls and having them work and learn with us. In this way we should be forced to learn more rapidly, systematically, and thoroughly, and they might get some idea of the usefulness and beauty of biology, carpentry, hygiene, interior decoration, accounting, and all the rest.* We should be very ignorant, but only good teachers know how ignorant they are. There would be no courses: the questions would be pursued as soon as possible after they arose. One difficulty would be the authentic liberalization of the practical, but I think it could be done. Another question would be that of time and finances. We could take on pupils only as a serious specialty: they could depend on it, we'd learn them or kill them. And we could not afford to do that without substantial payment. Only fairly rich parents could pay us well, yet they would be interested in College Boards, and hostile to most of my basic ideas; besides, we are much less interested in the children of the rich than in the children of the poor, or of impoverished gentlefolk. And my only connections with parents are through a school where my reputation, never substantial, will be completely demolished by the publication of these pages. Still, we may be able, some day, somehow, to work it out.

Meanwhile, such pedagogical instincts as we have do not suffer from lack of exercise. Domestication is artificial, and one has to discover and remember just how one is interfering with the life of an animal, learn from the mother, and teach as well as one can. I'd like to launch into a discussion of the teaching of small animals in relation to the teaching of those other small

*An invaluable text, and a model in the liberalization of the immediate and practical, would be Know Ross County, a geological survey for schools by our County Agricultural Agent, Fred R. Keeler.

animals, children, but as every parent will be convinced, I don't know enough about it. All I dare say is that the interest and fun are endless, and that the parallels are remarkable.

Of simplicity.

To me, Henry David Thoreau is one of the greatest and most interesting of men, but I think that if his life and work are to be fully appreciated, and made most useful today, they will have to be given much thought in relation to new experience, much more than I have had available for the purpose. I am therefore very hesitant in speaking of this colossal spirit. However, one can't live a life like this, and finish a book like this, without considering, in relation to them, a few questions that have been best raised by Thoreau.

With what I take to be Thoreau's first purpose, "to live deep and suck all the marrow out of life," I don't see how anyone can have anything but the most complete sympathy and admiration. In the clarity with which he saw his end, the re-assertion of the human spirit, and in the courage with which he pursued it, in his own way, this man was one of the great of the earth. As a seer, in the sense that we all have to be seers, to see, to think, in order to live deeply, he was like a normal man in an asylum for the blind.

Beyond this, his ideas and behavior seem to me less generally applicable, less worthy of unqualified admiration. Like all moralists, he was inclined to forget the glorious variety of human nature. In some lives, for living deep and sucking the marrow, solitude and inaction must be strictly limited, or become dangerous. Some characters are much better equipped for wide travel and worldly intercourse than for close observation of a village, its inhabitants, and its natural environs. Women, money, and machinery may be hard to dominate in the interests of deep living, to dissolve, as it were, in immediate and satisfying experience, but they are with us, willy-nilly, and I for one have occasionally found them extremely pleasant, and even deepening! The self-styled "reactionaries" to agricultural, pre-industrial democracy, are bad enough, without trying to go back beyond the invention of coins, back even beyond the creation of Eve!

Finally, take his major principle of simplicity, of simplifying life in order to get the most out of it. I don't know anyone, myself included, who in this respect can afford to be smug and flippant, who can dismiss too quickly this immortal challenge from the banks of Walden Pond. Everyone, if he has the courage and clarity, can simplify further to his own advantage. However, if we are unable or unwilling to give up society, action, women, money, and machinery so completely, it seems to me that simplicity of life is by no means generally and absolutely good as a technique.

As an humble instance, I imagine that many of our friends think of Peggy and me as living "the simple life." It should be obvious by this time that our lives are anything but simple, that they are immeasurably more complex than they were when we were single people living on good salaries in cities. As a means to seeing, to living deep, we may be said to be trying to develop and integrate what "faculties" we have, and this means many more activities, many more relations between them, much more complexity. If a man trusts and accepts society in flux more than I do, he can more easily simplify his own activities. This was evidently easier a hundred years ago, and it may be easier, in another hundred years, than it is right now. In a sense, the complexity of a farm, and of a farmer's life, are necessitated by the absence both of an erstwhile expansion, and of a future integration, of society. If I am unwilling to accept total dependence on an urban complex, unwilling to make the eliminations made by

Thoreau, unable to become a part of a socialized rural complex, and fearful of becoming the cast-off rural wreck of a machine age, like our "hill-billies," I damned well have to scheme and sweat in order to *keep* my activities sufficiently complex to get me what I want.

Besides, it seems to me that there is a good deal of cant thought and talked about simplicity in life, character, art, society, and everything else. Some day, somewhere, I am going to take a couple of cracks at it.

However, I must admit that Thoreau has given me food for a good deal of not wholly pleasant thought. Social organizations, gadgets, machinery, automobiles, tractors, stocks, bonds, electric companies, advertisements, newspapers, magazines, and all the rest: I trust them as I would an adder fanged.

Ad astra.

For some years, now, I have been almost obsessed by the old idea that human life is essentially theatrical, and that it can be both more intelligible and more fun if it is seen as such. I have mentioned this in connection with our friends in town, in the Little Theater. Here I want to suggest that a workman is more effective and happier if he dramatizes himself as such, that all social relations can be given style by a sense of the theater, that all play of people in groups is amateur theater of one kind or another, and finally, that even solitary pleasures may be called theatrical, in that they consist in contemplation of the world as a stage, and in enrichment of the individual for the parts he may play thereon.

I have acquired this obsession first because I am an American. That is, my ancestors were of many different nationalities and stations in life; my parents moved from place to place, and from milieu to milieu, relatively little, but almost enough to conform to the American tradition of experiment and home-lessness; and finally, I myself have moved about enough, and worked and played with enough different kinds of people, to make quite reasonable, and familiar to me, the rather stagger-ing question: "What *are* you, anyway?" Add some time in Eu-rope, where most people are cast into parts when they are born, and are trained to play them with resignation and style. Add finally literary curiosity, and perhaps a certain amount of the typically American slowness of development. The result is this obsession—and a very entertaining one I have found it, too.

Naturally, as it has grown, I have seen more and more people as case-histories of unconscious and attempted acting, and have become increasingly aware of theatrical demands made upon myself, and of my equipment, or lack of equipment, to meet them. However, this is hardly the place for that most de-lightful form of talk, theatrical reminiscence.

In the same way, I have noticed in my vagrant reading more and more discussions of this subject. It occupies some of the most brilliant of George Santayana's *Soliloquies in England*. It is one of the principal obsessions of William Butler Yeats, whose treatment of it is mystical, astrological, and very diffi-cult, but exciting. There are some good passages on it in L. H. Myers' novel, *The Root and the Flower*. The psychological depths of the subject have been plumbed by C. J. Jung, who seems to have used first the term "persona." It occupies a whole chapter in Aldous Huxley's *Eyeless in Gaza,* and several pages in the same writer's essay, "Writers and Readers," in *The Olive Tree*. In this essay Huxley uses Jules de Gaultier's useful and amusing term, "the bovaric angle," for the relation between internal "reality" and the "persona."

I can't explore the whole fascinating subject here, but I must note a few ideas, from observation and from this reading, that I have found useful, and that have stuck. The basic problem is that of uniting, or forming an effective alliance between, the soul, or the ego's response to the unconscious, and the persona, or the ego's response to external reality. In other words, the problem is the reduction of the bovaric angle. This can be done in either direction, that is, either by absorbing the persona into the self, or soul, or by projecting the self completely into the persona. The former method is that of romantic and egoistic self-expression and phantasy; the latter method is that of imitation, discipline, and action.

The dangers of the former method seem to be not only the encouragement of phantasy, loss of touch with reality, and ineffective action, but also a certain amorphousness and banality of personality. Despite all the great romantic "personalities," it may be said that without imitation, discipline, and action, personality does not exist. The great advantage of this method is that it permits the growth of the new.

The dangers of the other course are equally clear. The unconscious and the soul can be neglected at peril, and unless they find that the persona is a sympathetic and flexible medium, they will go their own way. The advantages of this course are in strength, effectiveness, interest, "character." Obviously, both methods must be used, with due attention paid both to the peculiar possibilities and necessities of the individual and to his probable environments and fortunes in life.

On the whole I should say that Americans, including myself, are now less in need of psychoanalytical, mystical, glandular, and astrological explorations and diagnoses of their inner selves, and of most kinds of "progressive education," than of some shrewd type-casting, hero worship, imitation, and discipline. If appropriate personae are hard to find in a time of flux, when external fortunes are so hard to predict, they are also especially necessary to inculcate the toughness necessary for survival. Reactionaries will be quicker than radicals to recognize this fact, and will be more shameless in their distortion of history, biography, and even anthropology. As an American radical, I find John Reed too childish for hagiography, and Lenin, Trotsky, and Stalin largely irrelevant. I promise myself to explore further Plutarch, the French Revolution, the First American Revolution, and the Dictionary of American Biography. I suspect that if some radicals don't make honest and effective use of these figures, quickly, reactionaries will turn them into hideous but effective caricatures. They will plunder and distort even Emerson, Whitman, Thoreau, Jefferson, and Jackson. For seventy years they have tried to use as their patron saint a frontier lawyer from Illinois whose actions, words, and suspenders do not seem to me to lend themselves very well to their purpose.

Speculations such as these may be interesting in themselves, but they are more interesting in their application to special cases. This brings us to the relevant point of this section, which is simply this: as an immigrant farmer, talking to others of similar temper, who may be thinking of doing similar things, I am quite unable to offer an adequate persona or even a hero, simply because I have been unable to find one for myself. The need is more acute than it may seem to be, and I consider the references I have given here to Santayana, Yeats, Myers, Jung, Huxley, and Gaultier quite as pertinent and quite as practical as references to pamphlets on chicken diseases, or on sheep breeding, which are easier to find.

It would be easy for us to play the eccentric country lady and gentleman, the last remnants of families gone to seed, running quickly through their financial and moral inheritance. For this rôle, we really ought to be regionalists and reactionaries, with shotguns and a pew in church, and even my radicalism and agnosticism might be dismissed as eccentricities, though less attractive than others. Plenty of people try to force us into this rôle, but I need hardly say that we don't find it congenial.

A good rôle demands an appreciative public and fellow-players with whom one can exchange whispered asides, but almost no one appreciates the rôle we are trying slowly to create. This fact reflects on the intelligence of no one, because even Thoreau, I imagine, did not create his part out of a void; and without models, we find it hard to understand our new rôle ourselves.

In fact, I find myself thinking less of classical and American heroes than of a picture by James Thurber. In this picture, at the base of a strange mountain, there squats, miserably nude, as though he had been ejected from a locker room without his clothes, a little man. On the flank of the mountain there lies sprawling, also nude, but with greater pleasure, a gargantuan female figure, a Maillol sculpture gone wrong, a laundress deified. She stares at the little man wild-eyed, and points, with a colossal flipper, towards a sky full of diseased stars. The title is: "Ad Astra."

Ad terram.

Illnesses have been mentioned in these pages. I don't intend to record them in detail, because I think physical pain is sheer waste and boredom. The pains of love, for instance, can be turned into poetry and fiction, but from my point of view, a pain in the anus remains a pain in the anus. When I suffer, I yell and swear; when Peggy suffers, she cries like a baby; and we are both fairly shameless about it.

Still, disease is such a dangerous enemy that I think a few facts have to be included, in the interests of realism and sober reflection. We are both fairly young, and in fairly good health, but in our first two years we had to spend about five hundred dollars on doctors, dentists, medicines, and hospital. In this period I lost three or four weeks, and Peggy three or four months. Peggy has suffered from hay fever, asthma, an allergic skin complaint, colds, intestinal grippe, and the painful loss of four teeth. I have suffered—a good deal less—from piles and their removal, sore throats, rheumatism in the hands and arms, two attacks of trench mouth, minor intestinal troubles, and the fairly easy loss of four teeth. I spent ten days in the hospital here, and Peggy has had to be taken twice to Columbus and once to Cleveland, to specialists. Peggy is now on a severe diet, and is giving herself serums once or twice a week. We are both now faced with the elaborate and expensive reconstruction of our mouths.

We are most lucky in our physician, who has all the virtues and none of the defects of the general practitioner, and is an old friend to boot. I wish he would start a Coöperative hospital.

How will it be, we sometimes wonder, when working all day indoors and outdoors, and coming in tired, but not too tired, and eating and drinking with gusto, and laughing, and going to bed in very love, are memories only?

But we only think such things when we are tired. The worst effect of fatigue is not among the obvious ones: exasperation at getting so little done, the limitation of one's activities, dependence on hired help, and so on. Its worst effect is its insidious corruption of all one's attitudes and emotions, of all one's waking hours. When I am too tired, I sink very low in my way; when Peggy is tired, she sinks very low in hers. We always drag each other down, and then, although we are as fond of each other as most married people seem to be, we could richly enjoy boiling each other in oil. In time, we may learn and accept the limits of our strength.

A typical effect of fatigue is worry. When we are exhausted, we find it easy to imagine all the bad things that must be happening to the animals and the land, it is easy to magnify the long chances we are taking, and to imagine ourselves sinking through illness, bankruptcy, and dependence, to death. Another cure of worry, besides rest, is the relaxed, alert fatalism that is slowly created by the very nature of farm work. Here, if anywhere, with these mighty enemies and allies, one can learn to feel: "if it be not now, yet it will come: the readiness is all." And so of death.

On a farm, death is always waiting to touch, gently and decisively, the trees and animals and plants that one cares for. Hardly a month can go by, without something dying, and having to be chopped down, and rooted up, or buried. Sometimes the pain is surprisingly, disproportionately keen, as when Peggy's dog was killed; but often, as when chickens die, there is no pain at all: simply a dull ache over one's incompetence to face this enemy. In any case, there is that moment when a small life gasps a little, or squirms a little, and is gone . . . so completely gone that one wonders whether it ever really existed. There is that extraordinary, that marvelous little organism, but it is already becoming cold and stiff, and if one does not go quickly to the tool room, and get the shovel, there will be maggots. These are life, too, and in certain places, or from certain perspectives, dignified and useful, but not on a farm.

The familiarity of this experience removes much of its pain. Death seems closer here than in a city, and less nasty. One seems to hear of more people living to sound old age, and then dying peacefully and naturally. Not long ago, one of our neighbors, a very old woman, died quietly, almost happily, and was buried, simply, by her neighbors and friends. Even the violent deaths of men in the woods, or in machinery, or from shotguns, or bulls, or lightning, seem less horrible, more natural than violent deaths in the cities. Here, the hazards are more obvious, and at best, every living thing has its season, and its end.

Or if more philosophy is required, it is not hard, here, to keep calm, and look at this thing, long enough to see the whole of life on this planet as the flowering of a little garden, the checkered and sanguinary flourishing of a little farm, in one brief spring, summer, and autumn, between two winters, the first without beginning and the second without end. It is not hard to see it all as a little accident, a brief improvisation, a folk song between silences, but as more than enough to send the chance listener in happy awe through the rest of his life to his grave.

But we are only men, busy, groping, hungry, and loving, and often, at these times, old hurts ache, and both heart and philosophy are weak. Often enough, I turn these little burying jobs over to James, hoping that he is not thinking, as I am, of slower, more pretentious burials. Sometimes there is something weak and frightened in me that keeps fighting, in my mind and guts, this transience, this waste of life and beauty, something that cries out in fear for the lives of my friends and comrades, and of my ally.

Then I think of the Homeric sailors, who wept, unashamed, and had their fires and games, and then ate well, and drank, and slept, beside their little boats, in the starlight. And then I pick up the limp and stiffening small body of a lamb, or puppy, and get the shovel, and do what I have to do, and go about my work.